Dr. M. J. Benton

THE PENGUIN
HISTORICAL ATLAS
OF DINOSAURS

PENGUIN BOOKS

Published by the Penguin Group
Penguin Books Ltd, 27 Wrights Lane, London W8 5TZ, England
Penguin Books USA Inc., 375 Hudson Street, New York, NY 10014, USA
Penguin Books Australia Ltd, Ringwood, Victoria, Australia
Penguin Books Canada Ltd, 10 Alcorn Avenue, Toronto, Ontario, Canada M4V 3B2
Penguin Books (NZ) Ltd, 182–190 Wairau Road, Auckland 10, New Zealand

Penguin Books Ltd, Registered Offices: Harmondsworth, Middlesex, England

First published 1996
1 3 5 7 9 10 8 6 4 2

Text copyright © Dr. M. J. Benton, 1996
Design and maps copyright © Swanston Publishing Limited, 1996
All rights reserved

The moral right of the author has been asserted

Printed and bound in Great Britain by The Bath Press, Avon

ISBN 0–14–0–51336–1

Foreword

Dinosaurs are the most popular extinct animals. Yet the dinosaurs were only a small part of the huge array of animals that once lived on the Earth, and there were many other groups of extraordinary creatures that were not dinosaurs. Life arose some 3500 million years (Myr) ago as simple single-celled microscopic organisms, perhaps something like bacteria. More complex organisms, big enough to see with the naked eye, arose only 800-1000 Myr ago, and the first animals with skeletons or shells are known from 550 Myr ago. Fishes date back to 520-530 Myr ago, and the first backboned animals on land came about 380 Myr. Dinosaurs arose about 230 Myr ago, and they ruled the Earth until their famous extinction 65 Myr ago, after which the mammals rose to prominence, and finally the first humans perhaps 5 Myr ago at most. Dinosaurs were not the oldest prehistoric animals by any means: indeed they were relative latecomers.

The Penguin Historical Atlas of Dinosaurs traces the history of the Dinosauria, their origins, diversification, and extinction. For the first time, it has been possible to plot detailed information on maps at all scales, from global and continental maps showing the distributions of major groups, to locality maps of dig sites. The main themes of the book are to illustrate the great success and diversity of the dinosaurs, to show how major dinosaurian groups came and went during their long reign on land, and to show how the evolution of dinosaurs related to major changes in the physical environment such as continental drift, and to changes in plant life and in other animal groups. An additional theme is historical: the first discoveries of dinosaurs, and their later study, over the course of nearly 200 years, have run in parallel with major changes in the ways in which scientists have viewed Earth history, time, evolution, and the life of the past. Dinosaurs are definitely not just for children!

I am extremely grateful to Malcolm Swanston for the inspiration and plan of this book, and to Andrea Fairbrass and Charlotte Taylor, and to other members of the Swanston Publishing team, for making the texts and illustrations into a book. I am also indebted to the generations of dinosaur palaeontologists whose work I have relied on for the material in this book.

Michael J. Benton

Bristol, 1996

Contents

I: How to Understand Dinosaurs

Dinosaurs were the largest animals ever to walk the earth, but there were many smaller species. Dinosaurs were some of the most successful animals that have ever lived.

Dinosaurs are often regarded as being synonymous with failure. Everyone thinks of extinction when the word 'dinosaur' is mentioned, and indeed advertisers use the term to mean some product or company that is too big and cumbersome to survive in the modern world. Perhaps, these people would change their minds if they returned to the Cretaceous period, and found themselves facing a hungry *Tyrannosaurus* or *Deinonychus*!

It is easier to make a case that dinosaurs were some of the most successful animals ever to have lived on the earth. Firstly, dinosaurs include the largest land animals of all time. Many of the sauropods, the long-necked plant-eaters, reached lengths of 20–30 metres, and weights of 50 tonnes or more. The largest living land animals, elephants, rarely exceed a body weight of 5 tonnes, and the largest extinct mammals, giant rhinoceroses that lived in Mongolia 25 million years ago, were only twice that size. A second argument for dinosaur success is that they dominated the terrestrial scene for 160 million years. Mammals, the group that includes humans, have only dominated the earth for the past 65 million years, so we have another 100 million years to go before we can claim to match the success of the dinosaurs. A third argument for dinosaur success is their great diversity, perhaps more than 1000 species, and the fact that they occupied every continent on earth, from the equators to the poles.

But surely, the dinosaur-hater says, dinosaurs are a failure because they went extinct in the end; they are no longer around. That argument is based only on our present standpoint. We are looking at vast spans of geological time behind us, and no doubt vast spans of evolutionary time in the future, from a mere instant in the whole history of the earth, or of the universe. Why should success be judged only on the tiny sample of plants and animals that happen to exist at a point 10,000 years after the beginning of the Holocene Epoch? Besides dinosaurs are said to be extinct—but are they?

The giant carnivorous theropod Tyrannosaurus *chases a herd of duckbilled dinosaurs. The duckbills were large enough, 5–10 metres in length, but Tyrannosaurus was a giant, the largest meat-eating animal ever on land. In general, dinosaurs were ten times the size of mammals. The smallest mammal is a tiny shrew, only a few centimetres long; the smallest dinosaur was* Compsognathus, *the size of a turkey. The largest living mammal, an African elephant, weighs 5–7 tonnes; the largest dinosaurs were ten times that size.*

Dinosaur Diversity

Dinosaurs are not a random assortment of giant and horrible extinct animals. The group Dinosauria is a real evolutionary unit. This means that all dinosaurs evolved from a single common ancestor, and that it is possible to trace the exact pattern of evolution within the group. The family tree of dinosaurs is reconstruct-

ed by studying their anatomy, and by searching for unusual features that may be shared by two or more species.

All dinosaurs share a number of specialisations that are not known in any other reptiles. Most of these special features of their anatomy arose because from the start dinosaurs stood upright on their hind legs. They were bipeds, as humans and birds are, but their limbs were also tucked beneath their bodies, while most reptiles are sprawlers. The ancestors of dinosaurs were sprawlers too, with their legs extending sideways, and bending down at the knees and elbows. The upright posture changed the orientation of all the joints (see left).

So far, about 1500 dinosaur species have been named, but many of these are not entirely convincing. Some were based on incomplete specimens, sometimes just an odd bone or tooth, and others turned out to be the same as species that had already been named. So, for example, the well-known sauropod *Brontosaurus* was named in 1878, but it was noticed later that it was the same as *Apatosaurus*, which had been named in 1877. The name *Apatosaurus* has precedence, and it must be used in preference to the later name *Brontosaurus*. The true number of named dinosaur species is much smaller than 1500.

Usually, people refer only to the genus names of dinosaurs, rather than the species. In fact, the only well known dinosaur species is *Tyrannosaurus rex*. The species *rex* is one of the species of the genus *Tyrannosaurus*, perhaps the only one. It has been estimated that just over 500 genuine dinosaur genera have been named so far. [A genus (plural, genera) is a group that contains one or more closely related species. Genus names are given in italics, and they have an initial capital letter.] The total of 500 is probably a very low estimate of the true diversity of dinosaurs in the past, and perhaps there genuinely were many thousands of genera during their long reign on the earth. At any time, several dozen, or several hundred, dinosaurs lived worldwide, but each genus lasted for 5–20 million years before it evolved into something else, or became extinct.

The Order Dinosauria ('terrible reptiles') falls into a number of subgroupings, each of them corresponding to a major split that happened during the evolution of the group. Soon after the origin of the dinosaurs, more than 230 million years ago, the group split into two, the suborders Saurischia ('reptile hip') and Ornithischia ('bird hip'). Saurischians are distinguished by features of their skulls, backbones, and arms, while ornithischians have a specialised set of hip bones, where the pubis, the front bone, runs back parallel to the ischium.

The suborder Saurischia includes two infraorders, the Theropoda ('beast foot') and Sauropodomorpha ('reptile'). Theropods are all the meat-eating dinosaurs, from tiny *Coelophysis* (see pages 74–75) to giant *Tyrannosaurus* (see pages 132–133). Sauropodomorphs are the long-necked plant-eaters, from the large prosauropods like *Plateosaurus* (see pages 76–77) to the giant sauropods like *Apatosaurus*, *Diplodocus*, and *Brachiosaurus* (see pages 98–99, 102–103).

The suborder Ornithischia includes only plant-eating forms, divided into five groups on the basis of major features of their limbs, heads, and armour, the Ornithopoda ('bird foot'), Ceratopsia ('horned face'), Pachycephalosauria ('thick-headed reptiles'), Stegosauria ('plated reptiles'),

Members of the Order Dinosauria are characterised by many specialised features that set them apart from all other reptiles. Dinosaurs stood fully upright on their hind legs, which were tucked right under their bodies. This upright posture caused major changes in the orientation of the various leg joints. The ankle and knee ceased to be complex angled joints, as in the sprawling ancestors of the dinosaurs, and they became simple back and forwards hinges. The hip joint also changed, so that the head of the thigh bone developed a ball-like shape which fitted firmly into a deepened socket on the side of the hip bones. Interestingly, humans have many of these features, because mammals also have the upright posture seen in dinosaurs, but, of course, humans evolved quite independently of the dinosaurs.

and Ankylosauria ('articulated reptiles'). The ornithopods were all bipeds, animals like *Iguanodon* (see pages 118–119) and the duckbills (see pages 126–133). The ceratopsians were mainly quadrupeds of the Late Cretaceous, animals like *Protoceratops* (see pages 126–127) and *Triceratops* (see pages 132–133), with horns on their faces and bony neck shields. The pachycephalosaurs were rarer bipeds, with extraordinarily thick bone in their skulls roofs. The two armoured groups, the ankylosaurs and stegosaurs were abundant in places during the Jurassic and Cretaceous (see pages 94–95, 102–103, 118–119, 130–133).

What of the birds? Careful analysis of the characters of dinosaur skeletons shows that birds are simply feathered dinosaurs. The oldest bird, *Archaeopteryx*, from the Late Jurassic of Germany, has the skeleton of a small theropod dinosaur. It has a long bony tail, strong fingers with claws on its hands, and sharp teeth in its jaws. Birds are living dinosaurs.

Dinosaur Days

Dinosaurs are not the oldest fossils. In fact, when the entire history of life is considered (see pages 48–49), dinosaurs are relatively modern. The earth was formed about 4600 million years (Myr.) ago, and the oldest fossils of simple living things date from 3500 Myr. ago. The oldest vertebrates, backboned animals, are simple fishes from Cambrian rocks, dated about 510 Myr. Vertebrates moved on to land during the Late Devonian, perhaps 375 Myr. ago, and the oldest reptiles are known from the Late Carboniferous, 320 Myr. ago. Dinosaurs arose 230 Myr. ago, during the Triassic period, and they ruled the earth from the Late Triassic, through the Jurassic period, and to the end of the Cretaceous period, 65 Myr. ago, when they died out.

The divisions of geological time, the Triassic, Jurassic, and Cretaceous periods, during which the dinosaurs ruled the earth, are grouped together in the Mesozoic Era (see pages 18–19). During this time, dinosaurs fed on a variety of plants. At first, the plants would have had a primitive aspect to our eyes— ferns, seed ferns, horsetails, cycads, and conifers. But, during the Cretaceous, the flowering plants (angiosperms) rose to prominence, and the last dinosaurs lived in rather incongruous association with modern-looking roses, oaks, and magnolias (see pages 132–133).

Flitting among the Mesozoic plants were dragonflies, flies, and beetles. Social insects, such as wasps, bees, ants, and termites, evolved in the Cretaceous, some of them closely adapted to pollinate and feed

Pterosaurs (above), *the Mesozoic flying reptiles, were close relatives of the dinosaurs. Pterosaurs ranged from small forms which were the size of pigeons, to the giant* Pteranodon, *with a 7–metre wingspan, and* Quetzalcoatlus, *with a colossal 12–metre wingspan. These giants were much bigger than any bird, and their aerodynamics were more like those of a glider than any other known flying animal. Pterosaurs ruled the skies until the end of the Cretaceous period, when most of the ruling reptiles died out.*

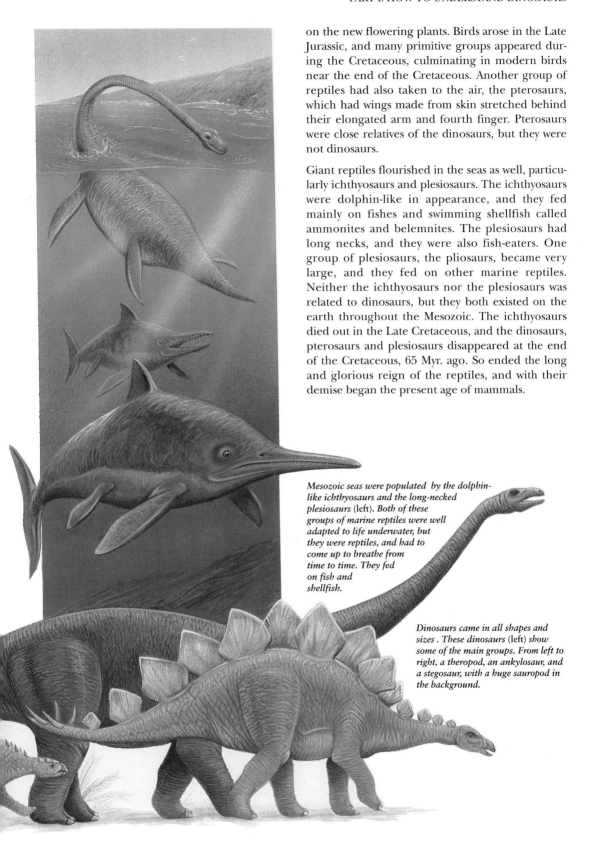

on the new flowering plants. Birds arose in the Late Jurassic, and many primitive groups appeared during the Cretaceous, culminating in modern birds near the end of the Cretaceous. Another group of reptiles had also taken to the air, the pterosaurs, which had wings made from skin stretched behind their elongated arm and fourth finger. Pterosaurs were close relatives of the dinosaurs, but they were not dinosaurs.

Giant reptiles flourished in the seas as well, particularly ichthyosaurs and plesiosaurs. The ichthyosaurs were dolphin-like in appearance, and they fed mainly on fishes and swimming shellfish called ammonites and belemnites. The plesiosaurs had long necks, and they were also fish-eaters. One group of plesiosaurs, the pliosaurs, became very large, and they fed on other marine reptiles. Neither the ichthyosaurs nor the plesiosaurs was related to dinosaurs, but they both existed on the earth throughout the Mesozoic. The ichthyosaurs died out in the Late Cretaceous, and the dinosaurs, pterosaurs and plesiosaurs disappeared at the end of the Cretaceous, 65 Myr. ago. So ended the long and glorious reign of the reptiles, and with their demise began the present age of mammals.

Mesozoic seas were populated by the dolphin-like ichthyosaurs and the long-necked plesiosaurs (left). Both of these groups of marine reptiles were well adapted to life underwater, but they were reptiles, and had to come up to breathe from time to time. They fed on fish and shellfish.

Dinosaurs came in all shapes and sizes. These dinosaurs (left) show some of the main groups. From left to right, a theropod, an ankylosaur, and a stegosaur, with a huge sauropod in the background.

"If then they are neither the bones of Horse, Oxen, nor Elephants, as I am strongly persuaded they are not, upon comparison, and from their likes found in Churches, it remains, that (notwithstanding their extravagant magnitude) they must have been the bones of Men or Women..."
Robert Plot

In his book The Natural History of Oxfordshire'(1676), *Robert Plot gave a detailed account of a huge broken bone (below) which had been unearthed in an Oxfordshire quarry. He identified it correctly as an ancient bone, and as the lower end of a thigh bone. Little did he know that he had in his hands the fist dinosaur bone ever to be described. The specimen is now lost.*

The Earliest Discoveries

Dinosaur bones must first have been found in ancient times, but no record was kept. In the Medieval period, philosophers commented on fossil shells and sharks' teeth which they had seen, and they debated the origin of these strange stones. Were such stony petrifactions in any way related to modern shells and fishes, or were they simply odd pebbles that happened to look like the remains of plants and animals? A popular view was that fossils were 'sports of nature' formed in the rocks by plastic forces. The first dinosaur bone to be described was found while this debate continued to rage in the 17th century, and the line of argument followed by its describer is revealing.

Robert Plot, Professor of 'Chymistry' at the University of Oxford was known to be preparing a book on the 'Natural History of Oxfordshire', and local naturalists sent him unusual specimens to examine. These included a weighty item that had been collected in a shallow limestone quarry at Cornwell in north Oxfordshire. Plot illustrated the specimen in a figure that also contained illustrations of numerous other strangely-shaped stones, some of which he interpreted as preserved kidneys, hearts, and even as the feet of humans. His interpretation of the rock from Cornwell was, however, altogether different.

Plot saw that the specimen looked like a bone. It had a broken end which was circular, and seemed to have a hollow core which was full of sand. The fractured surface round the core showed clear porous patterns exactly as in bone, and the shape of the opposite end, with its two large rounded processes, was just like the knee end of a thigh bone. Despite the seemingly mystical interpretations that Plot gave to many of the other stones illustrated in his book, he seemed to have no doubts about this giant bone. Going through a discussion of the kind of creature that could have produced such a monster bone, Plot stated that it came from an animal that was larger than an ox or horse, and considered the possibility that it might have come from an elephant brought to Britain by the Romans. He ruled out that possibility since the bone was even bigger than that of an elephant.

Plot's final decision was that the Cornwell bone came from a giant man or woman. He referred to mythical, historical, and biblical authority in support of this interpretation ('there were giants in those days'). After a promising discussion, Plot's final decision to identify the bone as human might seem perverse to us, but recall that no-one at the time had an inkling of the former existence of dinosaurs and other extraordinary animals in the history of the earth. Indeed, there was no acceptance of the idea that some plants and animals might have become extinct, since that would imply that God had made a mistake.

Plot's bone can be identified from his illustration as the lower end of the thigh bone of *Megalosaurus*, a dinosaur that is now well known from the Middle Jurassic of Oxfordshire. The specimen is now lost, but there is a final twist in the tale. The same bone was illustrated again in 1763 by R. Brookes, and he named it *Scrotum humanum* in honour of its appearance. This is the first named dinosaur, although unfortunately the name has never been used seriously.

The Idea of Extinction

Extinction seems a very obvious aspect of nature to us now, but in the

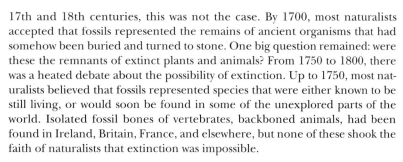

Baron Georges Cuvier (1769–1832) was the first comparative anatomist. He used detailed comparisons of the bones of extinct and living animals such as the tooth of an extinct mastodon (below), to show that fossil forms were distinct, and he used the same methods to reconstruct the appearances of ancient animals.

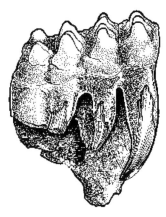

17th and 18th centuries, this was not the case. By 1700, most naturalists accepted that fossils represented the remains of ancient organisms that had somehow been buried and turned to stone. One big question remained: were these the remnants of extinct plants and animals? From 1750 to 1800, there was a heated debate about the possibility of extinction. Up to 1750, most naturalists believed that fossils represented species that were either known to be still living, or would soon be found in some of the unexplored parts of the world. Isolated fossil bones of vertebrates, backboned animals, had been found in Ireland, Britain, France, and elsewhere, but none of these shook the faith of naturalists that extinction was impossible.

Everything changed in the 1750s, when explorers in North America began to dig up the remains of elephants (mastodons and mammoths), and sent some of the bones to Paris and London. There, distinguished anatomists and naturalists, such as William Hunter, Georges Louis Leclerc, and Comte de Buffon, debated the specimens. As more specimens were found, and more of the Americas were explored, it became clear that these were remains of recently-extinct forms. Baron Georges Cuvier of the Muséum d'Histoire Naturelle in Paris was instrumental in clarifying this question. He showed, in a series of books and papers from 1796 onwards that the fossil elephants, and giant mammal bones from various parts of the world, represented extinct species.

The Vastness of Geological Time

The mammoths, mastodons, and giant ground sloths of the late 18th century were obviously not very ancient fossils. The bones were still in good condition, and the skeletons were often quite complete. Geologists accepted that some of these animals had perhaps died out only a few thousand years ago.

Other fossils seemed more ancient, the bone or shell material was filled with crystalline minerals, often calcite, quartz, or iron oxide, and these minerals must have taken some time to enter the pores and solidify. In many cases, the fossils appeared to come from plants and animals with no obvious living relatives. Geologists debated the antiquity of the earth but the outlines of what had happened in the past became clear only from 1800 onwards.

The first breakthrough came in the writings of James Hutton (1726–1797), a Scottish agriculturalist and naturalist. In his *Theory of the Earth* (1795), Hutton argued that the earth was enormously ancient, basing his argument on his observations of modern-day processes. Hutton observed the rates of accumulation of sediments, and the rates of erosion by mountain streams in his native Scotland, and he believed that such processes must always have been equally slow. He then compared modern rates of processes with the piles of

The skeleton of a giant sloth from Argentina. Here was an animal that was obviously extinct, but showed similarities to living animals of the areas, the smaller tree sloths. The skeleton shown here caused a sensation when it reached Europe, sent from Argentina in 1788 by a Dominican friar. It was exhibited in the Royal Collection in Madrid, and Cuvier obtained engravings of the skeleton, which he named Megatherium in 1796.

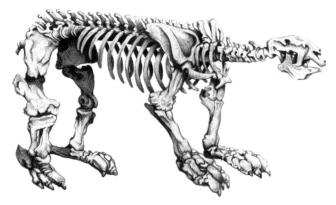

William Smith (1769–1839) found that fossils occurred in predictable assemblages and in a predictable order in the rocks. This is the basis of the geological time scale.

rocks he saw in cliffs around Scotland, and the depth and ruggedness of the mountains. These features must have taken millions of years, to form. Hutton's work was popularised by James Playfair, and the idea of an ancient earth spread rapidly among naturalists and philosophers.

The antiquity of the earth was a theme developed by later geologists, among them Charles Lyell (1797–1875). In his highly influential *Principles of Geology* (1830), Lyell called on examples from all over Europe to show how landscapes had changed, coasts had moved vast distances, and fossils were found abundantly everywhere. Lyell especially championed Hutton's application of modern processes to the interpretation of the past, the principle of uniformitarianism, or 'the present is the key to the past'. Between 1795 and 1830, practical field geologists had begun to produce geological maps, and to divide up the sedimentary rocks according to their fossils. This marked the beginning of geology as a science, and particularly the beginning of the modern international geological time scale (see pages 18–19).

Making Sense of the First Dinosaurs

The first dinosaur discoveries, from 1824 to 1840, began to paint a portrait of an astonishing fauna of giant reptiles in the Mesozoic. The dinosaurs, found first in England, and then elsewhere in Europe and, after 1850, on every other continent (see pages 22–29), showed early naturalists that whole assemblages of plants and animals, quite unlike anything now living, had existed in the past. How were palaeontologists to interpret these extraordinary giant bones that seemed to be quite unlike any living animal?

Georges Cuvier provided the method, in the new science of comparative anatomy. He carried out a painstaking comparison of every bone of a fossil form, and noted similarities and differences between equivalent elements in the skeletons of a variety of extinct and living forms. He found that the shapes of bones indicated the purposes for which they were used and the relationships of the animals in question. By the 1820s, Cuvier had honed his skills in comparative anatomy to such a pitch of perfection that it was said he could identify any animal from a single bone, and that he could reconstruct any unknown fossil form from a single bone.

When William Buckland and Gideon Mantell were trying to interpret the first dinosaur remains in England in the early 1820s, they sought Cuvier's help. He was as puzzled as any about the identity of the bones (see pages 22–23). Eventually, these palaeontologists realised that they had specimens of some kind of large reptile, and they believed that *Megalosaurus* and *Iguanodon* were in fact giant lizards. Richard Owen overthrew that idea when, in 1842, he argued that the dinosaurs were bulky quadrupedal reptiles that had many advanced mammal-like features (see page 23), and the true appearance of dinosaurs was arrived at only when Joseph Leidy described the first complete skeleton of a hadrosaur in 1858 (see page 24). How were palaeontologists to account for the origin and disappearance of the dinosaurs?

Evolution

The final link came with the theory of evolution by natural selection, proposed by Charles Darwin in 1859. Darwin had come to this theory after a long voyage of discovery in the early 1830s, both literally and figuratively (see

Charles Darwin (1809–1882), often reckoned to be the most influential scientist of all time. His explanation of evolution by natural selection not only changed the way biologists and palaeontologists viewed nature, but it also changed the way everyone thought about their place in the world. No longer was the world unchanging and simply created by God for human enjoyment. No longer was Homo sapiens *the pinnacle of creation. People had to get used to the idea that we stand in the midst of vast spans of evolutionary change, and there is no reason to assume that humans are in any way special, in any way an endpoint.*

pages 20–21). He set sail as gentleman naturalist on board the survey ship *Beagle* with the generally-held views of his day, that life had been created, and that species did not change. He came back a convinced evolutionist. Darwin did not propose the idea of evolution: this view had been championed by distinguished French naturalists, such as Buffon and Lamarck, in the 18th century. Their idea was that species were not fixed, and that they could change through time, although they were unclear about how this happened. That is all evolution means, literally 'unrolling' or change.

Palaeontologists throughout Europe had also been gathering evidence that pointed to some kind of change, or progression of life through time. They found simple fossils in older rocks, then fishes, then reptiles, then mammals, showing, they thought, some kind of sequence of development. Whether this progression meant that there had been a series of separate creations and extinctions, or whether ancient fossil forms had somehow changed, was debated in a general way.

During the voyage of the *Beagle*, Darwin saw and collected numerous examples of fossil mammals from South America, which confirmed that the fossil forms resembled mammals still living there. If life had been created, why should there be apparent evidence of some relationship between extinct and living animals in one part of the world? When he visited the Galapagos islands, Darwin saw that the plants and animals there were like those on the South American mainland. Why should that be, if life had been created? He saw also that the tortoises and finches on each of the dozen or so islands in the Galapagos group looked very similar to each other, and yet he observed that there were clear differences from island to island. When he showed his collection of bird skins to an ornithologist on his return, he was assured that the finches were all different species. Darwin had to admit the impossible; species were not immutable: if species were not permanent, then that meant they could evolve and split. This meant that God had not created all life in one act. All life could have evolved from a single common ancestor, and the vastness of geological time was already available for this process to take place.

The final brick was fitted into Darwin's shocking new edifice, when he read Thomas Malthus' *An essay on the principle of population*. This showed how human populations always breed faster than the increase in available food, and that certain processes must come into play to maintain the correct level of population. Darwin saw that this idea applied to animals and plants which all produce too many young to survive, and, in general, only the strongest live. The features that enable them to survive (bigger teeth, stronger legs, brighter feathers) must be inherited in some way, and they are passed on to their offspring. In time, the make-up of the whole population may change, or evolve, in the direction of the features that most promote survival at the time. This is natural selection.

Looking at the immense diversity of life today, and looking back over the history of life, admittedly only patchily known in the 1830s, Darwin could see a single principle at work. Species could evolve and split. In time those species themselves could change further and countless millions of years, a single population of simple organisms could have diversified into many species. A group like the dinosaurs could have arisen from a single ancestor which, if well adapted to the prevailing conditions, would survive and multiply. Over time, as new opportunities presented themselves, and with the vast potential for reproduction and variation, any kind of evolutionary change could be imagined: increase in body size, change from bipedalism to quadrupedalism, change in diet, or change in habitat.

From Organism to Fossil

Dinosaur fossils constitute just a tiny sample of all the dinosaurs that ever lived.

Dinosaurs are preserved as fossils. A fossil is literally anything that has been 'dug up', and when the term was first used, in the 17th century, fossils included the remains of ancient organisms, curiously shaped stones, and even potatoes. In modern definitions, there are two categories of fossils. Body fossils are the typical kinds, the shells or bones of ancient creatures, and leaves and trunks of ancient plants, and trace fossils are indications of the movements and other activities of animals, such as tracks and burrows.

Dinosaur body fossils are usually skeletons or isolated bones, although rarer examples include casts of the skin. Dinosaur trace fossils are isolated footprints and tracks of several, or many, footprints (see pages 104–105). Dinosaur fossils are uncommon in some ways: you can't walk out of your house and expect to find one in the nearest quarry or sea cliff. However, thanks to the efforts of hundreds of collectors around the world, there are now thousands of dinosaur bones and skeletons in museums, and dinosaur footprints are known from hundreds of localities. How did these fossils form?

The formation of a body fossil involves a number of steps, and the best way to imagine these is to think of a series of filters that tend to prevent preservation. The study of all the processes that go on between the death of a plant or animal, and its eventual discovery as a fossil is taphonomy. The main taphonomic filters can be arranged in sequence:

(1) The first filter is scavenging and decay. After a dinosaur died, its carcass would have lain on the ground where it fell, or in the river or lake where it drowned. On land, several processes can reduce even a large skeleton to dust within months. First, the flesh disappears rapidly, partly scavenged by meat-eating animals, and partly eaten away by decay microbes. If the climate is hot, and the dinosaur died in a desert area, the carcass might be covered with sand, which prevents scavenging. In rare cases, the conditions have allowed the bodily fluids to dry out without decay of the flesh, and the carcass has been mummified. If the dinosaur carcass was submerged under water, the flesh is equally likely to decay, or to be scavenged by flesh-stripping fishes and other animals. Whether on land or under water, the carcass is rapidly reduced to a pile of bones.

The processes of decay might end there in many cases, but even hard skeletal elements like bones are not entirely safe. In hot climates, and in situations where the bones are buried in certain kinds of soils, the bones may become weathered. Weathering is a combination of chemical and physical processes that break down hard materials, and bones can be cracked and reduced to shards under prolonged exposure to a hot sun.

(2) The second taphonomic filter is transport. The skeleton, or the few bones that were scattered by scavengers, may be picked up in a flash flood after monsoonal rains, or a river may change its course, and the bones are whipped along in a fast-flowing stream. On land, the bones could also be rolled around by strong winds. Transport by water, or by air, causes further physical damage, abrasion, wearing away of the delicate parts of the bones. Fresh bone is supple and stands up well to physical tumbling and rolling, but bones that have lost their protein content, perhaps during the decay and weathering

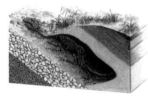

processes, become rounded and broken rapidly during transport. Many dinosaur bonebeds were created by river transport, and the bones are piled up in one place where they hit a sandbank, or where the river slowed down at a bend.

(3) The third taphonomic filter is burial. Carcasses or bones may be buried on land, or under water, at an early stage. Often, however, the bones have been weathered and transported first. Burial under mud or sand often preserves the bones from further damage. However, further destruction can occur. Bones can crack and become flattened under the weight of sediment piled up above them. After burial, water moving through the sediments can bring in dissolved chemicals that infiltrate the pores of the bones. Generally these chemicals are inert, and simply crystallise in the spaces, but sometimes, if they are acidic, they can cause damage to the bones.

(4) The fourth taphonomic filter is fossilization. During the early phases of burial, usually spanning the first few hundred thousand years, the sediments turn into rock by the loss of pore waters, and the cementing of grains into a hard immovable structure. The enclosed dinosaur bones also fossilize. All the spaces become filled with minerals, calcite (calcium carbonate), iron-containing minerals, and the like, and the mineral component of the bone, apatite (calcium phosphate), recrystallises to some extent, and may modify its composition by taking up ions from the pore waters.

(5) The final taphonomic filter is uplift, erosion, and discovery. At some point, the ancient sedimentary rocks may become exposed on the surface of the earth, and then they are subject to modern erosion by wind and rain. Most dinosaur bones are probably lost in the rapid degradation of badlands, in North America, Central Asia, and other parts of the world. In some cases, however, a palaeontologist, or keen fossil collector, chances on a specimen, and collects it. Even then, the specimen may be lost, or it may suffer damage.

After this catalogue of potential disasters, it may seem miraculous that there are so many dinosaur skeletons in our museums! However, it is important to remember that the dinosaurs lived on the earth for 160 million years and, during that time, billions of individuals must have lived and died. If they had all been preserved in perfect condition, the whole surface of the earth would be covered today by a pile of dinosaur skeletons several kilometres deep.

Sequence of stages of fossilization (above). A carcass is scavenged, and decays to a skeleton, which is then buried.

A mummified dinosaur from the Late Cretaceous of Canada (right). Soft parts of dinosaurs are only very rarely preserved. In certain hot desert-like conditions, however, a dinosaur carcass could be buried in sand, and the hot dry air allowed its body fluids to escape rapidly without decay of the flesh.

Telling Geological Time

Dinosaurs are dated according to the international time scale, based on a combination of fossil and radiometric evidence.

The earth formed about 4600 million years ago, the first simple life appeared 3500 million years ago, the fishes arose more than 500 million years ago, dinosaurs appeared 230 million years ago, and humans a mere five million years ago. These vast spans of time have been discovered only in the past two hundred years of geological research, first by the development of techniques of relative dating, and then by the techniques of absolute dating.

The first efforts to subdivide these vast spans of time according to recognisable events were made in England at the turn of the 19th century by William Smith, a canal engineer. Smith observed that certain rock types, and certain assemblages of fossils, always occurred in a predictable order. In his map of the geology of England and Wales (1815), he established for the first time a practical system of stratigraphy, the foundation of the geological time scale. In particular, he named various divisions of the Mesozoic rocks of England, and showed that they could be identified by their characteristic fossils.

The first discoveries of dinosaurs in the 1820s and 1830s (see pages 22–23) began to paint a portrait of an astonishing fauna of giant reptiles in the Mesozoic. At the same time, giant sea reptiles, the ichthyosaurs and plesiosaurs, were being dug up in marine rocks in England and Germany, and with them were found abundant coiled shellfish, the ammonites. These new fossil finds were the final part of the jigsaw that laid the foundations of a modern understanding of the history of life.

The graph below shows the decay of a radioactive mineral through time. The half life is the time it takes for half the radioactive parent element to be converted to a different form, the daughter element. The decay follows an exponential curve, a pattern of halving of the quantity of the parent element in a fixed time. The half life is used in radiometric dating, since the time taken for decay of the whole volume of the parent radioactive element is infinite.

The work of William Smith, and the dramatic fossil discoveries of the palaeontologists, laid the foundations of stratigraphy, the study of geological history. Smith's observations provided a basis for the idea of relative dating, which is now critical in forming a true picture of the sequence of events in the past. At its simplest, you can look at a pile of sediments in a quarry or cliff, and determine that the lowest rocks are probably the oldest, the highest the youngest. Then, by using fossils, it is possible to correlate the rocks of one quarry with those of another nearby. Correlation is now routinely done on a worldwide scale, based on the observation that particular fossils, or assemblages of fossils, termed zone fossils, lived worldwide for a short span of time.

The framework of the international geological time scale (right) was drawn up from 1820 to 1870. The geological eras, Palaeozoic ('ancient life'), mesozoic ('middle life'), and cenozoic ('recent life') were established about 1840, and these eras were divided into periods, the periods into epochs, the epochs into zones, and even subzones.

The final step in fixing the global stratigraphic scheme is to establish absolute dates. This became possible about 1920, when scientists realised that certain radioactive elements decay at predictable rates. It is possible to work out the half-life of an element, the time it takes for half to decay into something else. By comparing the proportions of the radioactive parent, and the non-radioactive daughter element, a precise radiometric date can be established. These can be cross-checked by assessing dates from different radioactive elements from the same rock sample.

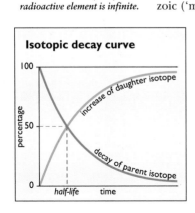

Isotopic decay curve

increase of daughter isotope

decay of parent isotope

percentage

100

50

0

half-life   time

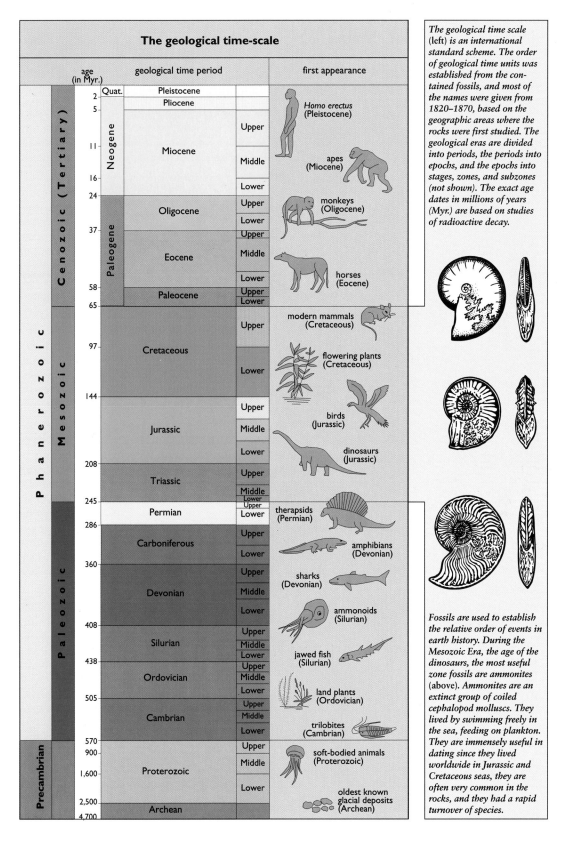

The geological time-scale

age (in Myr.)	geological time period			first appearance
2	Quat.	Pleistocene		*Homo erectus* (Pleistocene)
5		Pliocene		
11	Neogene	Miocene	Upper	
16			Middle	apes (Miocene)
			Lower	
24		Oligocene	Upper	monkeys (Oligocene)
37	Paleogene		Lower	
		Eocene	Upper	
			Middle	
58			Lower	horses (Eocene)
65		Paleocene	Upper	
			Lower	
		Cretaceous	Upper	modern mammals (Cretaceous)
97				flowering plants (Cretaceous)
			Lower	
144		Jurassic	Upper	birds (Jurassic)
			Middle	
			Lower	dinosaurs (Jurassic)
208		Triassic	Upper	
			Middle	
245			Lower	
		Permian	Upper	therapsids (Permian)
286			Lower	
		Carboniferous	Upper	amphibians (Devonian)
360			Lower	
		Devonian	Upper	sharks (Devonian)
			Middle	
408			Lower	ammonoids (Silurian)
		Silurian	Upper	
438			Middle	jawed fish (Silurian)
			Lower	
		Ordovician	Upper	
505			Middle	land plants (Ordovician)
			Lower	
		Cambrian	Upper	
			Middle	trilobites (Cambrian)
570			Lower	
900		Proterozoic	Upper	soft-bodied animals (Proterozoic)
1,600			Middle	
2,500			Lower	oldest known glacial deposits (Archean)
4,700		Archean		

Left column labels: Phanerozoic — Cenozoic (Tertiary), Mesozoic, Paleozoic; Precambrian

The geological time scale (left) is an international standard scheme. The order of geological time units was established from the contained fossils, and most of the names were given from 1820–1870, based on the geographic areas where the rocks were first studied. The geological eras are divided into periods, the periods into epochs, and the epochs into stages, zones, and subzones (not shown). The exact age dates in millions of years (Myr.) are based on studies of radioactive decay.

Fossils are used to establish the relative order of events in earth history. During the Mesozoic Era, the age of the dinosaurs, the most useful zone fossils are ammonites (above). Ammonites are an extinct group of coiled cephalopod molluscs. They lived by swimming freely in the sea, feeding on plankton. They are immensely useful in dating since they lived worldwide in Jurassic and Cretaceous seas, they are often very common in the rocks, and they had a rapid turnover of species.

Evolution

Life has evolved by a mixture of chance and natural selection, as Charles Darwin discovered.

Charles Darwin, pictured as a young man, when he developed his ideas on evolution. Darwin had begun his career as a medical student in Edinburgh, but he hated the crudity of surgical procedures at the time, and switched to a divinity course in Cambridge. He hoped to become a country parson, with leisure time to pursue his passionate interest in natural history. His life changed when he spent five years on board the Beagle in the early 1830s, studying plants, animals, and geology, especially in South America, and on some of the Pacific islands. On his return, he established a strong reputation both as a biologist and a geologist, based on his reports from the voyage of the Beagle. He spent the rest of his life, supported by his private income, writing a series of books which explained evolution, and which mark the beginning of the modern natural sciences.

Evolution occurs by natural selection, the 'survival of the fittest', as Charles Darwin proposed in his book *On the Origin of Species* in 1859. Populations are highly variable, and generally far more offspring are produced than can survive.

The core of Darwin's proposal was that evolution had occurred by natural selection. When Darwin visited South America, during the voyage of the *Beagle* in the early 1830s, he studied some of the unusual living mammals of the continent, the sloths, anteaters, and armadillos. He also collected fossil mammals from the same areas, and found that they were similar to the living mammals. Why, he asked, would this be the case if God had created life, and life could not change or evolve? Darwin realised that the fossils proved that the fossil mammals were related to the living forms by direct descent, that evolution had occurred.

On the Galapagos Islands, he famously studied the land tortoises and the finches (see page 15). He saw that there was a different subspecies or species on each island, and that these were obviously all related to each other. As a geologist, he knew that the islands had emerged from the sea as volcanoes in the not too distant past, and he could not avoid the conclusion that wanderers from South America had populated the new islands, and had evolved on them. This confirmed for him that species are not immutable—that they can change.

The third, and final, observation that Darwin made in the mid 1830s showed him the process by which 'evolution occurred, namely natural selection. There is a vast overproduction of young organisms, far more are born than can ever survive. It is possible to divide the theory of evolution by natural selection into several propositions, each of which can be proved in nature. These propositions are: (1) more young organisms are produced each generation than can survive into the next generation; (2) individuals within populations are variable; (3) only certain individuals within a population, bearing certain variations, will survive into the next generation; (4) those individuals that have special features enabling them to survive and reproduce (e.g. the ability to find more food, the ability to attract a mate, the ability to escape predators) will survive and reproduce; (5) many advantageous features of the surviving organisms are heritable, and will be passed on to future generations; (6) hence, through time, the make-up of a population, the range of adaptations it contains, will change, or evolve.

Variation within populations is also obvious. A particular pattern of banding in a snail may camouflage it and so save its life from a predatory bird. This is an adaptation, a feature that confers survival value. Natural selection shapes the adaptations of species. As environments change, the nature of selection continually changes. New species form when a parent population is divided by some barrier to reproduction, and the two halves of the population go their separate evolutionary ways. This important consequence, the generation of diversity, was an important element of Darwin's thinking, and it stemmed from his observations of the abundance of living and fossil species, and the links between species, both in time and in space, which he had observed in South America and on the Galapagos Islands.

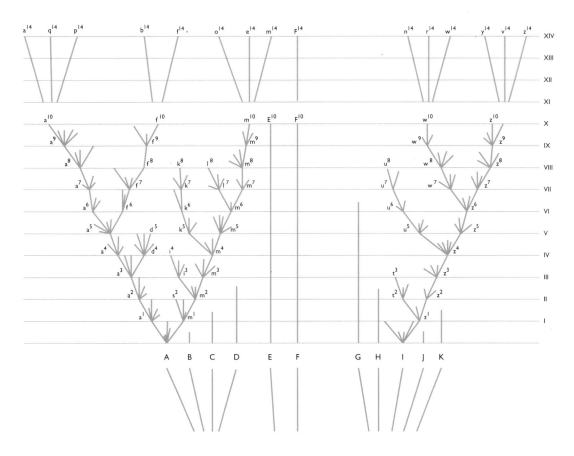

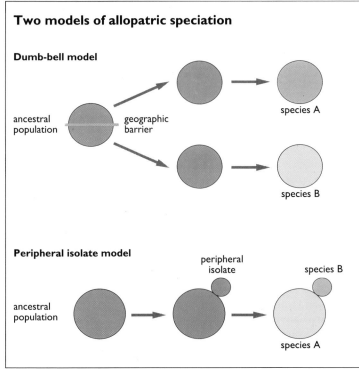

Two models of allopatric speciation

Dumb-bell model

ancestral population — geographic barrier — species A

species B

Peripheral isolate model

peripheral isolate

species B

ancestral population — species A

The first evolutionary tree (above) from Darwin's 'On the Origin of Species' (1859). In this diagram Darwin, shows how 11 species (A–K) may continue unchanged (E–G), die out without evolving (B–D, H, J, K), or form the basis for numerous speciation, or splitting events (A,I). Species A and I radiate, or evolve, and they are said to have been successful in evolutionary terms. Splitting between species is Darwin's explanation for the present huge diversity of life. A single species can split (left) by the establishment of a barrier to reproduction. Perhaps a mountain chain arose, or a river, or some part of the population became isolated. Natural selection continues acting on the two parts of the formerly united population, and they inevitably diverge. The split of the population can be roughly equal (top example), or unequal (below). Often a small part of a population settles on an island, and this peripheral isolate may evolve relatively rapidly if selection pressures are different on the island from those on the mainland.

First Finds

The first discoveries of dinosaurs revolutionised ideas about the history of the earth.

Dean William Buckland (1784–1856), Professor of Geology at the University of Oxford, and Dean of Christ Church (above). In the early 19th century it was not uncommon for churchmen to combine their careers with science. Around 1818 Buckland was shown some collections of bones and teeth of a large meat-eating reptile , but he could not identify the bones, and he showed them to Baron Georges Cuvier of Paris, the leading anatomist of his day, and to other experts. In the end, Buckland classified the animal as a giant reptile, probably a lizard, and he estimated it had been 40 feet (12 m) long in life.

Dinosaur bones were found hundreds of years ago, but the first printed record dates from 1676, when Robert Plot, the Professor of Chemistry at Oxford University published an illustration of a giant bone that had been found in a quarry at Cornwell, Oxfordshire (see page 12). Plot debated the identity of this huge specimen. He argued that it was too big to have come from an elephant (he had seen one in a travelling menagerie in Oxford), and in the end he decided that it must have come from an ancient giant man or woman. As evidence, he quoted the Bible: 'there were giants in those days'. The specimen is now lost, but Plot's illustration shows that it was the lower end of the thigh bone of the meat-eating dinosaur *Megalosaurus*.

This interpretation may seem laughable, but at that time there was an active debate about the nature of fossils, and many philosophers argued that they had been created in the rocks by 'plastic forces', and that perhaps God had placed them there when he created the earth. By the 18th century, geologists had settled the 'plastic forces' debate, and accepted that fossils represented ancient organisms. However, they were not so sure about extinction. In the 1750s, explorers in North America began to dig up the remains of fossil elephants (mastodons and mammoths). As more and more specimens were found, and as more of the Americas were explored, it became clear that these truly were remains of recently-extinct forms.

The first dinosaur discoveries in the 1820s and 1830s showed palaeontologists a glimpse of the extraordinary reptiles of the Mesozoic. Bones of a large meat-eating dinosaur came to light north of Oxford about 1818, and they were taken to William Buckland. Eventually, in 1824, Buckland published a description of the bones, and he stated that they came from a giant reptile which he named *Megalosaurus* ('big reptile'). This was the first dinosaur to be described.

At the same time, but independently, Gideon Mantell (1790–1852), a country physician in Sussex, was amassing large collections of Mesozoic fossils. During a visit to a patient near Cuckfield, so the story goes, his wife Mary, who had come with him, picked up some large teeth from a pile of road-builders' rubble. Mantell realised the teeth came from some large plant-eating animal, and when he sent them to Cuvier, the great French anatomist assured him that the animal must have been a rhinoceros. Then, Mantell compared the teeth with those of other modern animals in the Hunterian Museum in

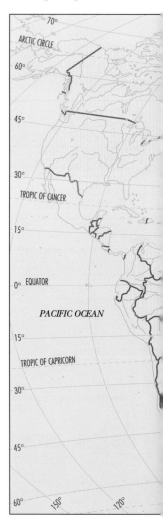

London, and a student there, Samuel Stutchbury, showed him that they were like the teeth of a modern plant-eating lizard, the iguana, only they were much bigger. So, Mantell described the second dinosaur, named by him *Iguanodon* ('iguana tooth') in 1825, basing his description on the teeth, and some other bones he had found since.

Richard Owen (1804–1892) surveyed the large Mesozoic reptiles, and he realised that *Megalosaurus, Iguanodon,* and the others, were not giant lizards. He called them dinosaurs ('terrible reptiles') for the first time in a publication in 1842, and in this, he named the fourth British dinosaur, *Cetiosaurus* ('whale reptile'), although he believed it was actually a giant crocodile.

The first discoveries of dinosaurs (below), from 1824 to 1850, came mainly from England, with a few other discoveries in Germany and France. At that time, most scientists lived in Europe, and large bones were generally brought to them by local quarrymen. Up to 1850, very few noted geologists actually went out looking for dinosaurs. The big expeditions would follow later in the century.

Owen argued that his new group, Dinosauria, consisted of advanced, almost mammal-like, animals. He had some evidence from the very limited range of bones known at that time: these showed that dinosaurs were reptiles, but they had very variable tooth shapes, depending on their diet. Owen believed that the dinosaurs were all quadrupeds, hefty rhinoceros-like animals. He was an anti-progressionist, which means that he opposed the idea that there had been progress in the history of life from simple to complex organisms (see page 15). Advanced mammal-like dinosaurs seemed to him to prove that reptiles had degenerated in some way since the Mesozoic.

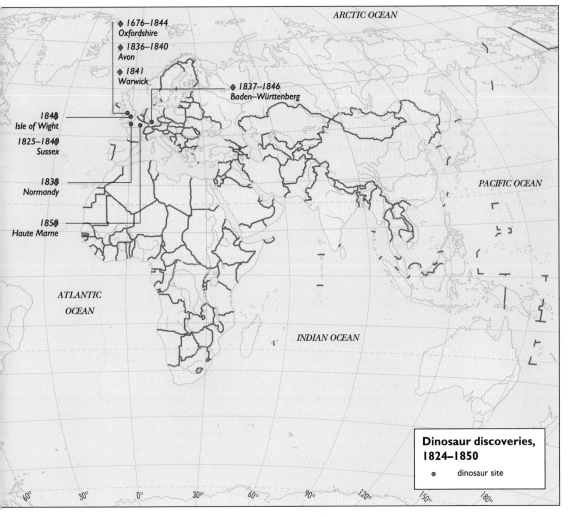

◆ 1676–1844
Oxfordshire

◆ 1836–1840
Avon

◆ 1841
Warwick

◆ 1837–1846
Baden–Württenberg

ARCTIC OCEAN

1848
Isle of Wight

1825–1848
Sussex

1838
Normandy

1850
Haute Marne

PACIFIC OCEAN

ATLANTIC
OCEAN

INDIAN OCEAN

**Dinosaur discoveries,
1824–1850**

● dinosaur site

The Heroic Age

Most of the famous dinosaurs from North America were collected in a frenzy of activity from 1870 to 1900, spurred by the intense rivalry between two men.

Edward Drinker Cope, probably the most prolific namer of reptiles, both living and fossil. During his research career he gave names to over 1000 new species, including fishes and mammals as well. The long-term rival of Othneil Charles Marsh, Cope was a brilliant man, but aggressive.

The first North American dinosaur skeleton, Hadrosaurus, *as mounted in the Philadelphia Academy of Natural Sciences by Joseph Leidy in the 1860s.*

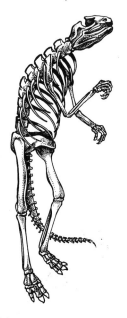

Dinosaur discoveries up to 1850 had relied on chance finds, usually by quarrymen, in Europe. These discoveries continued, with significant new finds in southern England, France, and Germany. Probably the most important dinosaur find in Europe between 1850 and 1900 was the mass grave of dozens of skeletons of *Iguanodon* found at Bernissart in Belgium. In 1877, coal miners working a deep shaft, more than 300m below the surface, came upon large bones in the roof of a cutting. These were reported to the local authorities, and a careful excavation followed. In the end, more than thirty complete skeletons were mapped and removed from the mine.

From 1850 to 1900, the main scene of dinosaur discoveries shifted to North America. Dinosaur footprints had been discovered in New England earlier that century, although they were thought to have been made by giant birds. The first dinosaur bones from North America were modest enough, a few teeth collected by an official geological survey team operating in Montana in the year 1855. They were described in 1856 by Joseph Leidy (1823–1891), Professor of Anatomy at the University of Pennsylvania. Small beginnings. However, two years later, Leidy was able to report a much more significant find, a nearly complete skeleton of a large plant-eating dinosaur from Haddonfield, New Jersey, which he named *Hadrosaurus*. Leidy realised that *Hadrosaurus* was related to *Iguanodon*, but it was younger in age. Most significant was the fact that the skeleton was more complete than anything yet known from Europe, and it proved for the first time that this dinosaur at least stood on its hind legs. Hitherto, dinosaurs had all been reconstructed as quadrupeds (see page 16).

The North American bone wars began in earnest in the 1870s. Edward Drinker Cope (1840–1897) was taught by Leidy in Philadelphia, and his interests spanned palaeontology and herpetology, the study of modern amphibians and reptiles. Othniel Charles Marsh (1831–1899) was also an enthusiastic palaeontologist, and he secured a position as Professor at Yale University through the patronage of his wealthy uncle George Peabody. These two men worked together in a friendly manner at first, both of them collecting and describing a range of fossil reptiles and mammals, first from the East Coast, and then from the new territories of the Midwest of the United States. Their rivalry began about 1870, because the ambitions of both men led to compete with each other for the best finds. Each of them

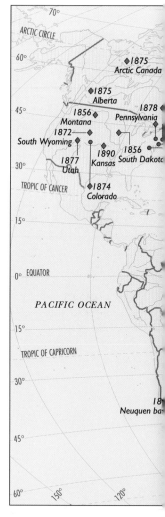

had enough money to buy fossils from local collectors, and to employ teams of excavators who operated in the Midwest. When word came through that a survey team had found some fossil bones, Cope's and Marsh's agents would gather a team of fieldworkers, and have them work day and night removing bones at speed. Some of these operations were in dangerous country, and the field men were armed against attack. At times, they worked through winter, removing bones in appalling conditions.

At the end of a field season, the dinosaur and mammal bones were loaded into boxcars and sent east by rail when possible, where Cope and Marsh fell on the packing cases, tearing them open, and describing the new dinosaurs and other wonderful beasts in haste. They rushed their manuscripts to their editors and published new dinosaur names as fast as the presses could roll. These papers by Cope and Marsh were brief and often unillustrated, such was their desire to outsmart each other. Thanks to Cope and Marsh we have all the famous North American dinosaurs—*Allosaurus, Apatosaurus, Brontosaurus, Camarasaurus, Camptosaurus, Ceratosaurus, Diplodocus, Stegosaurus, Triceratops* (see pages 100–103, 130–133). Inevitably, many of the dinosaur names established by Marsh and Cope have later turned out to be invalid because of duplication.

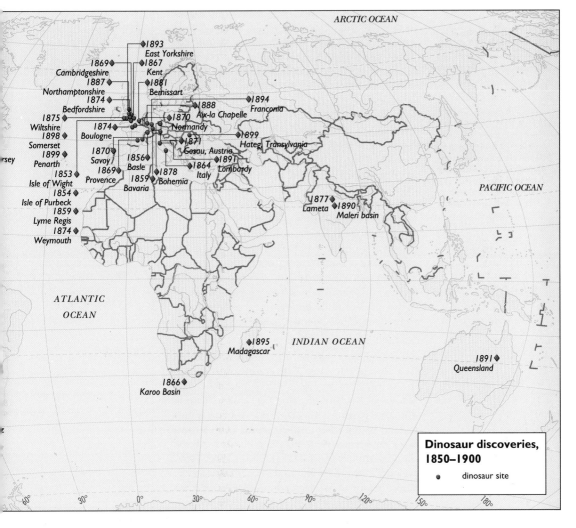

Dinosaur discoveries, 1850–1900

● dinosaur site

Worldwide Dinosaurs

Dinosaur-hunting expeditions from 1900 to 1950 opened up rich new territories to palaeontologists.

Charles H. Sternberg (1850–1943) began collecting dinosaurs for Edward Cope (see page 24), and he worked as a professional fossil collector for a number of institutions from 1876 into the early 20th century. He, and his sons George, Charles, and Levi, made important collections in the badlands of Alberta, and their specimens formed the nucleus of collections in Toronto and Ottawa, as well as in New York, and further afield.

The rivalry between Cope and Marsh from 1870 to 1900 had shown the richness of one major dinosaur-collecting area, in the American Midwest. Palaeontologists began looking further afield after 1900, driven both by the desire to know more, and also the increasing accessibility of parts of Asia, Africa, and South America. Collecting continued in the 'old' well-known dinosaur hunting grounds of Europe and North America, and important new finds were made. Collectors, such as Charles Sternberg, followed the dinosaur beds across the border from Montana and the Dakotas into Canada, and they found similarly rich dinosaur beds in Alberta about 1900. Further south, dinosaurs came to light in Texas, Arizona, and New Mexico. In Europe, collectors continued their efforts, and found dinosaurs at new localities in England, Belgium, and France, as well as in Scotland, Monaco, Romania (see pages 128–129), Poland, and Russia.

One of the biggest bone-hunting expeditions ever was mounted by German palaeontologists in what is now Tanzania, then German East Africa. Huge bones were found on a remote hill called Tendaguru by a geologist working for a prospecting company, deep within the African plains, and four days' march from the nearest seaport, Lindi. Huge sums of money were raised in Berlin, and a massive expedition operated at Tendaguru from 1909–12, led by Werner Janensch. On each visit, up to 500 local workmen were hired to do the digging. Bones were excavated, roughly protected with plaster, and carried by hand to Lindi for shipment to Berlin. Over 200 tonnes of bones were removed in this way, necessitating nearly 5000 trips between the site and the port, and it is estimated that 80,000 or more bones were collected. This huge effort yielded the complete skeletons of *Brachiosaurus*, *Elaphrosaurus*, *Kentrosaurus*, and others, in the Berlin Museum (see pages 98–99). Work in Africa also continued in Egypt, where dinosaurs were found by German palaeontologists in 1910–11.

Dinosaurs from Asia were found sporadically at first. Specimens from China were collected up to the 1920s by American, French, and Swedish priests and explorers. The first good Chinese dinosaur skeleton, *Mandschurosaurus*, was found in Manchuria, then under Russian control, and was sent back to St Petersburg in 1917. Asiatic dinosaurs really hit the headlines in 1922, 1923, and 1925, when an American expedition, led by Roy Chapman Andrews, and with Walter Granger as chief paleontologist, returned from

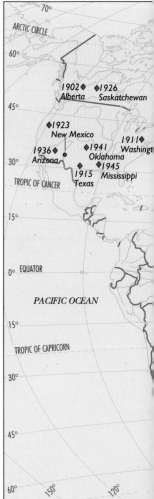

Mongolia with spectacular dinosaur specimens, skeletons of the small ceratopsian *Protoceratops* with nests containing eggs, the extraordinary slender meat-eaters, *Saurornithoides*, *Velociraptor*, and *Oviraptor*. The expedition had been sent out by the American Museum of Natural History in the hope of finding early human fossils, but it came back with evidence for one of the best dinosaur-hunting areas in the world (see pages 126–127). Major expeditions have continued to visit Mongolia since the 1920s, led by Russians, Poles, and Americans, collaborating with their Mongolian colleagues, and these have yielded numerous extraordinary finds.

More isolated discoveries were made up to 1950 in Russia, China, Egypt, Brazil, and Argentina. Initially, the work was stimulated by expeditions mounted mainly by European and North American institutions, keen both to acquire exotic specimens, and to expand their scientific reputations. Increasingly, during this time, countries in Asia, Africa, and South America established their own geological surveys, universities, and museums, and the focus shifted to locally-based experts. By 1950, dinosaurs had been found on every continent, except one, Antarctica, and more than 500 dinosaur species had been named. This was a dramatic rate of discovery when it is recalled that dinosaur hunting on a large scale began only after 1850.

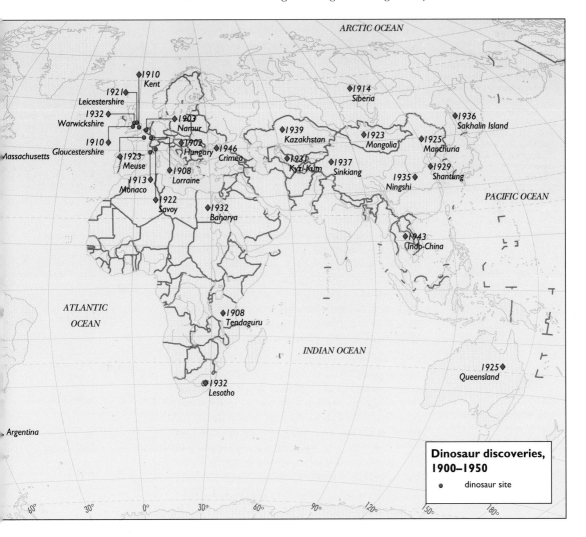

Dinosaur discoveries, 1900–1950

● dinosaur site

Latest Finds

Dinosaurs are still being described at the rate of one new species every three or four weeks.

Paul Sereno of the University of Chicago announces the discovery of the complete skeleton of a new theropod dinosaur, Afrovenator, collected during his 1994 expedition to Niger. This dinosaur proves land connections with North America during the Early Cretaceous, and it is only one of the new dinosaurs that are coming out of the relatively untapped dinosaur beds of North Africa.

Most research work on dinosaurs stopped during the Second World War, for obvious reasons. Research money was diverted to improving the technology of weapons, and large parts of the world that had produced dinosaurs before were out of bounds. After the war, very little work on dinosaurs was done until the 1960s. There were expeditions by the Russians to Mongolia in the 1950s, and these resulted in spectacular finds of the giant flesh-eater *Tarbosaurus* (see pages 126–127). French expeditions turned up dinosaurs in Portugal, Morocco, Tunisia, and Madagascar, and Chinese palaeontologists began to discover new dinosaur localities.

In the 1960s, Polish palaeontologists forged a link with their colleagues in Ulan Baatar, and a series of Polish-Mongolian expeditions took place. These expeditions were spectacularly successful, turning up more specimens of the classic Mongolian Late Cretaceous dinosaurs (see pages 126–127) *Protoceratops*, *Oviraptor*, *Velociraptor*, and *Tarbosaurus*, but also some new sauropod specimens, *Opisthocoelicaudia* and *Nemegtosaurus*, and an extraordinary specimen which has come to be known as the 'fighting dinosaurs'. This specimen preserves a skeleton of the lightweight flesheater *Velociraptor* with its claws seemingly locked into the bony headshield of a *Protoceratops*. Were the two fighting, *Protoceratops* perhaps defending its nest from a raid by *Velociraptor*, or is this merely a chance association of two skeletons? The Polish-Mongolian expeditions scored another coup in turning up dozens of beautiful skeletons of the tiny mammals that lived side by side with the dinosaurs. Previous expeditions had yielded a few mammal skeletons, but no-one had previously collected carefully enough to be able to bring such beautiful delicate fossils back to the lab. Dinosaur collecting has continued since the 1970s by Mongolian and Russian palaeontologists, and, in the 1990s, by a renewed collaboration between Mongolians and the American Museum of Natural History.

Dinosaur study began to expand seriously in China about 1965, with the establishment of several museums, and the training of a new generation of palaeontologists. Since then, dinosaurs have been found in Jurassic and Cretaceous rocks in China, and 100 or more species have been reported (see pages 94–95, 126–127). The rate of collection of new dinosaurs in China shows no sign of slowing down.

Dinosaurs had been found in Argentina and Brazil before, but young palaeontologists in both

Dramatic new discoveries of dinosaur eggs have been made in north China in the 1990s. Some localities are reported to contain thousands of eggs, preserved by some catastrophe that inundated a nesting ground. Some of the eggs contain unhatched embryo dinosaurs.

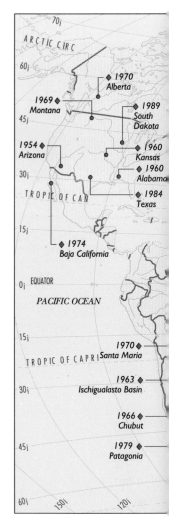

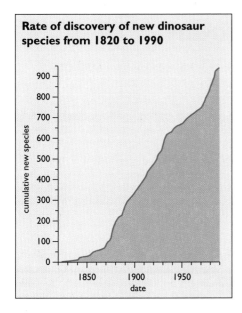

Rate of discovery of new dinosaur species from 1820 to 1990

countries began to find more material in the 1960s, and the pace has quickened especially in the 1990s, with dramatic new finds of the world's oldest dinosaurs in the Late Triassic of Argentina (see pages 70–71), and whole faunas of new forms from the Jurassic and Cretaceous (see pages 96–97, 124–125). The Late Cretaceous dinosaurs of South America in particular are important, since some forms are unique to that continent. In Africa, sporadic collecting has turned up dinosaurs from north to south, and east to west. Recent American expeditions to Niger and Morocco, and other parts of the Sahara, have shown that Cretaceous dinosaurs of Africa still had some affinities with North America, even though Africa was virtually an island by then (see pages 122–123).

The final piece of the jigsaw of dinosaur distribution was completed in the 1990s with the discovery of several dinosaurs in the Jurassic and Cretaceous of Antarctica. Dinosaurs are now known from every continent on earth.

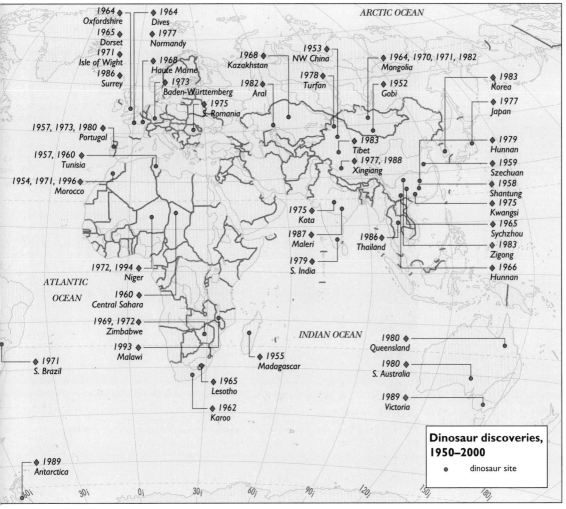

Dinosaur discoveries, 1950–2000

● dinosaur site

Digging up Dinosaurs Today

Dinosaurs are still found by trial and error, and dinosaur collecting remains a laborious business.

There are no secrets about collecting dinosaurs. Neither has modern technology improved collecting methods much since the 1880s (see pages 30–31). Dinosaur collectors still have to live rough often in remote parts of the world, and they still have to pace out vast areas of ground while prospecting. Collectors still have to engage in back-breaking digging and rock-breaking to clear the skeletons, and they still have the laborious job of protecting the bones, and carrying or dragging them out of the site by hand.

Modern technology has helped in the quality of camp life, so that food supplies and camping equipment are now generally better than conditions experienced by Cope and Marsh's collectors in the Wild West over a hundred years ago. Collectors now use all-terrain vehicles to bring in equipment, and to remove the massive bones. However, there are many times when a modern collector wishes he still had horses to pull specimens out, since trucks cannot drive up narrow mountain tracks or rocky gorges! Power tools can now be used to break up the rock lying on top of a skeleton, the overburden, and that may save some muscle work. Photography and film also allow the mod-

Right: *Excavating a dinosaur in the classic dinosaur-collecting country in the Badlands of Alberta, Canada. Here, erosion by heavy rainfalls prevents soil from developing, and the bare rock is exposed all the time. Here, the team begins to remove overburden from above a dinosaur skeleton. Collectors prospect up and down the gullies in search of bone shards which may lead them to a complete skeleton.*

Above: *Stages in excavating and plastering dinosaur bones before transport.*

ern collector to record what is going on much better than in former times. Modern collectors also have a finer appreciation of the context of their fossils, and they will usually have geologists on hand to record the sediments, and any evidence for ancient environments. They will also collect the bones more carefully, and they will try not to miss small specimens. Even in the 1950s and 1960s, some collecting in Mongolia and China was pretty crude: the main equipment consisted of bulldozers and high explosives! Techniques such as these destroy more than they preserve.

The first step in recovering a dinosaur skeleton is to find it. Collectors prospect by walking up and down exposed areas of rock which are known to be of the right geological age. The collector looks for clues, such as small shards of bone in stream beds, that might lead him back to a skeleton. When something is found, the collector may carry out some trial digs to try to work out whether there is a complete skeleton, or just some disarticulated fragments in the rock. If it is a complete skeleton, a team of five or six people usually moves in. The overburden is removed as rapidly as possible, and then the final centimetres of rock over the bone level are worked more carefully by hand. This is classic excavation, scraping and brushing away the debris until bones are exposed. Ideally, the whole skeleton is laid bare like this from above, and then the palaeontologists can map out and photograph the whole skeleton in context. The initial excavation phase, up to mapping, may take five or six people two weeks to complete.

After all the records have been made, the skeleton has to be removed for study. Dinosaur bones may be big, but they are not necessarily tough. In the heroic days of collecting (see pages 24–25), bones were levered up with crowbars, and thrown on to horse wagons. If a bone shattered, the collectors swept the fragments away, and tried another. About 1900, some collectors realised the terrible damage they were causing, and they tried strengthening bones with plaster. Gradually, they adopted a standard technique which has been used worldwide for as long as anyone can now remember. Sackcloth (burlap) is torn into strips, 5–10 cm wide, and these are passed through a watery plaster mixture. The strips are slapped over the bone, usually five or six layers thick; twice that for a large or awkward specimen. Before plastering, the bones, or groups of bones, are placed on pedestals by deep trenching all round. The shell of plastered cloth is wrapped low down the sides of the bones, and a little down the sides of the pedestals. Then, the collectors burrow away beneath the plastered bones, and tip the bone, and part of the pedestal, over. They can then scrape out as much rock from beneath the bones as possible, before plastering over the bottom. Each bone, or group of bones, is now entirely encased in plaster. Large specimens have thick beams of wood plastered across them to act as splints. Sometimes, the wooden splints are arranged to act as a kind of sledge.

If the excavation site is readily accessible, the plastered bone parcels are hauled on board. More usually, the bones have to be hauled out. Small ones are carried by hand, larger ones pulled by teams of people using ropes. The collectors have to build smooth roadways to haul out the large bone block sledges. Back in the laboratory, the plaster jacket is removed by the use of saws or other cutting equipment. The bones are handled carefully as they are cleaned up. Broken pieces are repaired, and the whole specimen is usually impregnated with a hardening agent that fills the pores and strengthens the bone. The skeleton may be stored for scientific description, or it may be prepared for display to the public.

Making Dinosaurs Live

It is difficult to make accurate restorations of dinosaurs from the starting materials of fossilized bones. A mix of science and art is essential.

"Any scene from deep time embodies a fundamental problem; it must make visible what is really invisible…it must make us 'virtual witnesses' to a scene that vanished long before there were any human beings to see it"
Martin Rudwick,
Scenes from Deep Time, 1992

Dinosaurs are dug out of the rock, and their huge bones are brought back for study in the back rooms and laboratories of museums and universities around the world. How do palaeontologists put the skeletons back together and make the dinosaurs live and breathe? The processes of excavation (see pages 30–31) inevitably involve a great loss of information. When the bones are moved, their exact context in the rock is lost, any associations with each other, or with other fossils that might indicate habitats and diets. In rare cases, some traces of soft tissues may be preserved, and that is hard to retain during excavation, Techniques today are as careful as they can be, and evidence can be preserved in film and photographs of the excavation. Also, the bones are treated carefully in the lab, and they are skilfully prepared to ensure that damage is minimised.

Things were very different when the first dinosaurs came to light. In those days, it was rare for the palaeontologists to be present when the bones were dug up. Usually they were chanced on by a quarryman, who might bring in only a few of the more complete bones to be identified. The rest of the skeleton could so easily be lost. University professors then did not generally get their hands dirty, and often they would not even visit the site from which the bones came, but would simply set about trying to interpret the fragments they had before them. Indeed, it has only been in the last fifty years that palaeontologists have made serious efforts during excavations to retain the tiny, often microscopic, fossils of small animals associated with the dinosaurs. They bag up sediment, and sieve it to retrieve tiny teeth and jaws, and these help to fill out the picture of ancient life.

Scientific drawing is an important part of the process. Palaeontologists themselves are often accomplished artists, and professional artists have a critical contribution to make, both in making accurate drawings of the bones and skeletons, and in helping in the process of bringing the bones to life by drawing whole animals, or reconstructed scenes.

Once the bones of a dinosaur skeleton have been prepared and conserved using consolidants, the palaeontologist seeks to identify the specimen. The bones of dinosaurs, like those of all other vertebrates, have very specific shapes, and it is usually not difficult to decide what each bone is: a left femur (thigh bone), a right humerus (upper arm bone), a particular element from the ankle, a neck vertebra. Usually, it is an easy matter to identify the group to which the dinosaur belongs, since the bones are very distinctive, and since so much information has been published about dinosaur discoveries since 1824. The femur or humerus of each major dinosaur group is quite different, and an expert can decide at once if he is dealing with a theropod, a stegosaur, or an ankylosaur. Even isolated teeth are enough to pin the group down. This can be done rapidly. Then the hard work begins to determine whether the new find belongs to a previously-described species. Here, the age of the rocks can narrow the search. If the fossil site is Middle Jurassic or Late Cretaceous in age, that will imply a very different array of dinosaurs, since those time units are separated by 100 Myr. or more, and a great deal of evolution took place during the interval.

Sometimes, very rarely, the new specimen turns out to be like some previously named dinosaurs, but not *the same* as anything yet known. Then the palaeontologist may start to become excited: is it a new species? If it is, the palaeontologist has the privilege of naming it, of choosing a generic and specific name that has not been applied to anything else, and of publishing a full and detailed account in a scientific journal. It is easy to make mistakes,

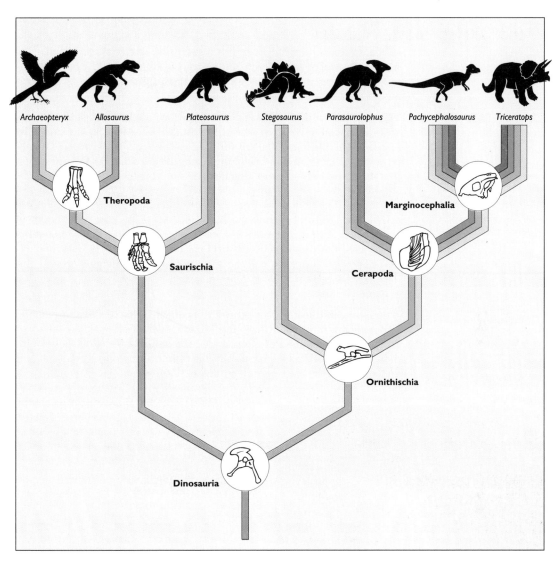

The evolution of dinosaurs (above) follows a pattern of splitting of the major groups. Palaeontologists can reconstruct the pattern of splitting, the evolutionary tree or phylogeny, by seeking shared derived characteristics, features that are unique to particular groups. So, all the dinosaurs share a particular hip structure. Within Dinosauria, all saurischians share particular kinds of hands and feet, and within Saurischia, the meat-eating theropods and birds share further specialised features of their hands and feet.

but the community of other palaeontologists will be very quick to inform the world at large if a mistake is made. This is a normal part of science, the constant checking and critical testing of new proposals, and it does not mean that scientists are any more or less unpleasant to their colleagues than any other group of professionals. It is important to be really certain that the new specimen is truly different from everything hitherto described before rushing into print with a new name. Despite the lessons of the past, that at least half of all the supposedly new dinosaur species names proposed turn out to be synonymous (the same as a species previously proposed), palaeontologists are ever hopeful. But, the naming of species, although important, is merely a question of semantics. How do palaeontologists interpret the old bones as living animals?

There are several ways of interpreting ancient organisms, some of them easily defensible, and others somewhat more speculative. This is the study of functional morphology, the interpretation of function from morphology, the shape and form of an animal. The main assumption behind functional morphology is that structures are adapted in some way, that they have evolved to be reasonably efficient at doing something in particular. So, an elephant's trunk has evolved to act as a grasping and sucking organ to allow the huge animal to reach the ground, and to gather food and drink. *Stegosaurus* presumably had large bony plates on its back for some purpose, and we might reasonably assume that the flick-knife toe claw of *Deinonychus* and *Velociraptor* was adapted for a very special purpose.

The bones of a fossil skeleton themselves can show directly how much movement was possible at a particular joint, and this is determined simply by manipulating the fossil bones, if they are well enough preserved. There may be muscle scars on the surface of the bone, and particular knobs and ridges, that show where the muscles attached, and how big they were. Muscle size is an indicator of strength, and this kind of observation can show how an animal moved. The bones show direct evidence of other soft structures that are not usually fossilised, such as the shape and size of the brain, the layout of major nerves and blood vessels, the location of certain glands, the size of the guts, and sometimes the presence of fleshy crests. These techniques, of interpreting joint movements, and the shapes of muscles, are fairly certain, and the clues can be read directly from the bones. Palaeontologists, however, like to go much further.

There are three main approaches to the study of functional morphology. First is comparison with living animals. If the fossil form belongs to a modern group, the palaeontologist can compare the bones of the fossil species with those of a modern representative. If there are no close living relatives, or if the living relatives are very different from the fossil species, then there may be problems. This is the case for the dinosaurs, which admittedly have living relatives in the birds, but there is not much to be gained by comparing a *Tyrannosaurus* skeleton with a modern sparrow or seagull. In the case of dinosaurs, modern *analogues* are used, that is living animals that may not be in any way related to dinosaurs, but which appear to have similar skeletons or similar parts of skeletons. So, for example, a palaeontologist studying the giant sauropods can compare his dinosaurs with modern elephants or rhinoceroses, since both groups are large heavy animals, and they share features of their pillar-like legs. When studying small lightly-built theropods, it may be helpful to compare them with modern ostriches, which have a similar body shape, and can offer clues about running speeds, balance, and feeding habits.

The second approach to studying functional morphology is to use mechanical models. The jaw of a dinosaur, for example, may be compared to a lever, and calculations may be made of the forces acting to close the jaw. Changes in the shapes of jaws in herbivorous and carnivorous dinosaurs can then be understood in terms of adaptations to achieve a stronger bite at the front of the mouth, or perhaps to evolve an efficient grinding and chewing system further back in the mouth. The layout of bones in the skull may be interpreted in terms of the stresses acting in dif-

Dinosaur bones are often put on show. There is a great deal of delicate work involved in mounting the hundreds of bones in a single skeleton for display. For practical reasons, museum technicians often make casts of the bones, perhaps using a tough lightweight material like fibreglass. This is good practice since it preserves the original from damage, and the casts are lighter and easier to manipulate than real fossil bones.

Stegosaurus, the famous plated reptile. For years, palaeontologists have speculated about the function of the plates. Were they for defence, temperature control, or for attracting mates? The temperature control function has been tested. Model arrangements of metal plates were tested in a wind tunnel, and the staggered arrangement seen in the dinosaur turned out to be the most efficient pattern for shedding heat in a slight breeze. In life, the plates were covered with skin which carried a rich supply of blood. When the animal became too hot, heat was radiated; when it was too cold, it could pick up heat through the plates by basking. The blood from the plates passed rapidly into the body, and transferred heat or cold. into the body core.

ferent directions in a hypothetical model of a box with holes. The limb joints may also be interpreted as engineering structures, designed to support weight and to move the dinosaur around. Simple calculations of bone strength allow the palaeontologist to work out that large sauropods could not have exceeded certain speeds or they would have broken their legs. Other features may be compared with engineering models of functions. The plates on the back of *Stegosaurus* (see pages 102–103) were tested against a model of cooling fins in a wind tunnel. The precise arrangement of plates seen in *Stegosaurus* turned out to be the most efficient for heat loss. This is not proof that *Stegosaurus* had evolved its back plates for that purpose, but it is suggestive evidence.

The third approach to functional morphology is to use supplementary information from the context of the fossil. Dinosaurs may be found in a variety of sedimentary settings, usually in the sediments of ancient rivers or lakes. There may be associated plants and animals that hint at their diet. There may be clear indicators in the rocks of the climate, features such as specific soil types, mudcracks, evidence for monsoonal rainfall, and the like. Some skeletons preserve remnants of stomach contents, and fossil dung, coprolites, are also known. There are even some specimens of dinosaur bones with tooth marks which show that the skeletons were scavenged after death.

Dinosaurs lived in communities in which some animals ate others, some specialised in eating particular plants, and others suffered from particular parasites. Just as today, organisms have always interacted in different ways with other organisms, and with the physical environment. The study of ancient modes of life and interactions is palaeoecology, and the focus of study may be on a single animal or on a whole community. It is possible to draw up lists of all the species of dinosaur, and other plants and animals that lived together, based on detailed excavations of single prolific sites. The modes of life of each species can be deduced from their bones and teeth, and then a diagram of who ate whom may be constructed, in other words, a food web. If there are enough specimens of some of the species, detailed measurements may show the presence of sexual dimorphism, that is, two sets of adult individuals, one presumably female, and the other male. Sometimes, juveniles are found, and these can show how the animal grew up.

All of the information from functional morphology and from palaeoecology is put together to produce a picture of an ancient dinosaur site as a living community. The palaeontologist can argue the case, based on his detective work, and then specialist artists, working for museums or for book publishers, can interpret the evidence as a colour painting, just like those in this book. The final stage, of making moving dinosaurs, has also reached a high pitch, with close collaborations between palaeontologists and artists going into spectacular museum displays of mechanical models, and computerised animation in films.

Were the Dinosaurs Warm-Blooded?

One of the constant questions palaeontologists have asked about dinosaurs is whether they were warm-blooded or not. How can such a question be answered without a time machine and a giant rectal thermometer? The debate began virtually with the invention of the group. In 1842, when

Richard Owen coined the name 'Dinosauria', he speculated that these giant reptiles had been rather mammal-like in their physiology, and very different from living reptiles, such as lizards and crocodiles. Owen clearly thought that dinosaurs were able to control their body temperature to some extent, and to keep it high.

Modern animals divide up into two main categories in terms of temperature control. The ectotherms, like fishes and reptiles, generally use only external means to control their body temperature. So, lizards bask on rocks to raise their temperatures, or hide in holes to cool down. The endotherms, like birds and mammals, can maintain warm body temperatures by internal means, by burning up food and by a complex feed-back mechanism that heats and cools the body to maintain temperature at a precise level. A change of only a few degrees can be critical, as in humans, whereas ectotherm body temperatures may vary by 20°C or more each day. The second distinction in modern animals is between poikilotherms and

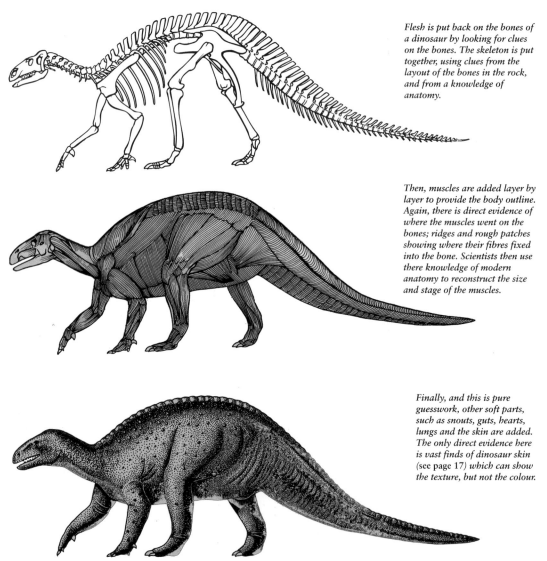

Flesh is put back on the bones of a dinosaur by looking for clues on the bones. The skeleton is put together, using clues from the layout of the bones in the rock, and from a knowledge of anatomy.

Then, muscles are added layer by layer to provide the body outline. Again, there is direct evidence of where the muscles went on the bones; ridges and rough patches showing where their fibres fixed into the bone. Scientists then use their knowledge of modern anatomy to reconstruct the size and stage of the muscles.

Finally, and this is pure guesswork, other soft parts, such as snouts, guts, hearts, lungs and the skin are added. The only direct evidence here is vast finds of dinosaur skin (see page 17) which can show the texture, but not the colour.

homeotherms. Poikilotherms have variable body temperatures, and homeotherms have constant body temperatures. Lizards are clearly poikilothermic ectotherms, and birds and mammals are generally endothermic homeotherms. But poikilotherm does not equal ectotherm, or homeotherm equal endotherm. The four terms are necessary because fishes are generally ectothermic homeotherms: their body temperature is constant, although controlled externally, since the temperature of the sea does not change much. Bats and hummingbirds are poikilothermic endotherms, since they can switch off their expensive heating system at night, or in winter. What were the dinosaurs?

Until 1970, dinosaurs were thought to be sluggish cold-blooded reptiles, in other words, ectothermic and poikilothermic, despite Owen's classic ideas. Then in 1970, Bob Bakker, at that time a graduate student at Yale University, marshalled a range of evidence that dinosaurs had been endothermic homeotherms. He noted these points, among others: (1) dinosaurs have complex bone structure with evidence of constant remodelling, a bone feature seen in modern mammals, but not in reptiles; (2) dinosaurs have an upright posture, as in modern mammals and birds; (3) dinosaurs evidently had active lifestyles, or at least the small theropods certainly did; (4) predator-prey ratios of dinosaurs show more in common with those of mammals than with those of modern reptiles; (5) dinosaurs are found in polar regions.

Bakker's collection of evidence shows some of the ingenuity palaeontologists must employ in their efforts to understand the past, drawing here on the physiology and anatomy of modern animals, bone histology, palaeoecology, and palaeogeography.

There was a furious debate about Bakker's proposals in the 1970s and 1980s. Most of his evidence was equivocal, and did not stand up to strong scrutiny. However, the observations noted above are all correct, and might suggest true endothermy. However, further study of bone structures has shown that the dinosaur and mammal pattern is associated with large size and fast growth rather than simply with endothermy. Upright posture does not necessitate endothermy, nor does an active lifestyle (think of insects or small lizards). Predator-prey ratios for dinosaurs do suggest that the predators were endothermic, but there are serious problems in trying to calculate such palaeoecological measures in a precise manner. Dinosaurs are found in regions that lay near the poles in the Jurassic and Cretaceous (see pages 120–121), but there were apparently no polar ice caps in the Mesozoic, and so conditions were not cold.

The consensus now is that dinosaurs were generally ectothermic homeotherms. That is, the large ones at least, were ectothermic, like modern reptiles, but because of their size they maintained a constant core body temperature. Large animals take a long time to cool down and warm up, and large size confers a property termed mass homeothermy. Some of the smaller predatory theropods might have been endothermic, however. Palaeontologists will never know the answer to these kinds of questions about the biology of dinosaurs, but it is possible to explore possibilities and probabilities in a defensible way. Also, such debates are great fun, even if, perhaps, especially since, they will probably never be resolved.

II: Before the Dinosaurs

Dinosaurs evolved after millions of years of evolution on the earth. Life arose 3500 million years ago in the sea, and marine animals became common 550 million years ago. The move on to land, and the origin of reptiles, preceded the origin of the dinosaurs by many millions of years.

A scene in Ediacara times, some 600 million years ago. The oldest large multicellular animals included sea pens, jellyfish and worms. The Ediacara fossils, found first in Australia, and then worldwide, document the first step in the dramatic expansion of life in the sea.

Dinosaurs are often seen as archaic animals, surely somewhere close to the origins of life. This is not at all the case: dinosaurs, like humans, came very late in the story of earth history. If the history of the earth is reduced to a single day, 24 hours, dinosaurs lived only in the last hour, and humans did not arise until the last two minutes.

The earth arose some 4600 million years ago, and at first it was molten and inhospitable to life. Any living thing that happened to arrive on that early earth would have burnt up immediately. Slowly, the planet cooled, and a crust of solid rocks began to form. Volcanic gases and water vapour formed a primitive atmosphere, and oceans began to accumulate when the temperatures had fallen low enough. By 4000 million years ago, there was rainfall and erosion on the surface of the earth, which is proved by the existence of ancient sedimentary rocks.

The first evidence for life is found in rocks dated as about 3500 million years old. There are two kinds of fossils that confirm the existence of life then. First are some tiny simple cells found in ancient cherts from Africa and Australia. The cells are preserved in the glassy silica-rich cherts as impressions, but they appear to show a cell wall of some kind. Also, evidence for early life exists in the form of stromatolites, irregular layered structures, some of them measuring several tens of centimetres in height. Stromatolites are formed today generally in highly saline warm waters by blue-green algae, or cyanophytes. The cyanophytes form a green slime on the seabed, and they survive by photosynthesising in sunlight. Fine particles of limy sand waft over the slime, and the cyanophytes grow up and spread out again. In time, a many-layered structure, a stromatolite, grows up. Fossil stromatolites preserved the characteristic sediment layers, the organic slime has long since disappeared.

More complex cells arose 1000 million years ago, and multicelled plants and animals a little later than that. True animals are known from about 600 million years ago, strange worm-like and coral-like creatures preserved in sandstones in Australia, and in many other parts of the world. The history of the earth, up to 550 million years ago, is generally assigned to one great division of geological time, the Precambrian, since this was a time of sparse evidences of life. The fossil record became much richer from 550 million years ago to the present, a time span known as the Phanerozoic, or 'abundant life'.

Life in the Sea

It is not clear why fossils became more abundant 550 million years ago. Perhaps life really was meagre and sparse up to that time, or perhaps all previous plants and animals were soft-bodied and so not very likely to leave fossils. Nevertheless, 550 million years ago, dozens of sea creatures evolved hard skeletons of some kind—shells, and external and internal skeletons. The origin of skeletons in so many different groups of animals, and all apparently at

Trilobites were the first complex animals. Trilobites are arthropods ('jointed limbs'), distant relatives of modern crabs, spiders, and insects. Trilobites ('three lobes') have three body portions from left to right, a middle body segment, and two side regions, and from front to back, a cephalon (head), thorax (body), and pygidium (tail). They lived on Palaeozoic sea floors, feeding on a variety of small prey.

the same time, is a mystery. The new skeletons were composed of a variety of minerals, including calcite and aragonite (both of which are forms of calcium carbonate) and apatite (calcium phosphate), and this increases the mystery of their simultaneous origin. It may be that some feature of the chemistry of the oceans changed at that time, or perhaps some new fiendish predators had appeared and only those organisms with skeletons survived.

Most striking among the first skeletonised organisms, those that lived from 550–500 million years ago, during the Cambrian period, were the trilobites. These were the first animals to see the world around them. Other animals before the Cambrian lacked eyes, but most trilobites had complex eyes, composed usually of many neatly arranged rows of calcite lenses. These compound eyes, just like the eyes of modern insects, must have meant that trilobites saw the world many times, each lens recording a slightly different image from the others around it. Typical trilobites had all-round vision, and this provided protection from predators. Trilobites were arthropods, animals with jointed external skeletons. Arthropods today include crabs, lobsters, spiders, and insects. Trilobites had three-part skeletons (hence the name, 'three-lobed'), consisting of a head shield, a jointed body portion that bore numerous legs underneath, and a tail cover.

Trilobites and other early arthropods are superbly preserved in the Burgess Shale, a deep-water marine deposit from the Mid Cambrian of British Columbia, Canada. Charles Walcott, Director of the Smithsonian Institution in Washington, is said to have discovered the first Burgess Shale specimen by accident in 1909, when he was out riding with his wife on a trail high in the Canadian Rockies. His horse kicked some loose pieces of shale, and he spotted a silvery impression of some ancient animal. He leapt down, and began turning over the thin black slabs with mounting excitement, and located some more fossils. The animals seemed to be preserved complete, with their delicate legs, and even many of their soft parts, in place. Original thin cuticles of arthropods were flattened to a thin film, but every detail was there.

Walcott mounted several expeditions to British Columbia, during which he collected thousands of fossils. Since then, many more specimens have been excavated, and they grace the collections of many museums. New techniques of study, such as X-rays and scanning electron microscope photographs, have revealed some of the astonishing detail of the Burgess Shale fossils. Since 1909, some 120 species of invertebrate animals (animals lacking a backbone) have been described from this one location. There are ancient sponges, molluscs, worms, and echinoderms, but the best-represented forms are the arthropods. Some are trilobites, others are close relatives of trilobites, but many belong to groups that were not known before. In the 1980s and 1990s, many more Burgess Shale-type faunas have come to light in the Cambrian of China, Greenland, and elsewhere in North America.

The first reefs arose in the Cambrian, but they were not made by corals, but by an unusual group of sponges, the archaeocyathans. These were cupshaped organisms, consisting of an outer and an inner porous stony sheet linked by cross-pieces. They grew on the shallow sea floor in tight clusters, and provided a home for other smaller animals.

Corals arose a little later, in the subsequent Ordovician period (500–440 million years ago). There were the two main groups, the rugose ('wrinkled') corals, with irregular folded patterns on the outside of the skeleton, and the tabulate ('flat') corals. Some rugose corals existed as solitary individuals, cone-shaped and sausage-shaped forms. Other rugose corals, and all the tabulates, formed colonies in which numerous separate animals lived in close association, each in its own little stony cell within a larger complex structure. The separate animals in these colonies were linked by living tissue strands by which they could communicate. Ordovician seas also abounded in jellyfish, sea anemones, and other relatives of the corals.

On the sea bed around the early reefs a variety of shelled animals, some mobile, and others fixed, fed on debris and on floating particles of organic matter. The brachiopods, sometimes called 'lamp shells', were the most abundant two-shelled animals on the seabed. They mainly lived attached to rocks and reefs, and fed by filter-feeding small particles of food from the sea-water. Brachiopods existed at that time in many shapes and sizes, and the group dominated the Palaeozoic Era (550–250 million years ago), but dwindled after that time. Molluscs, the other group of shelled animals, also occurred in the Cambrian and Ordovician, but they were much less abundant than the brachiopods. Molluscs include the bivalves, two-shelled forms such as oysters, clams, and mussels, gastropods, coiled snails and whelks, and cephalopods, swimming forms like squid and octopus. The early Palaeozoic molluscs were generally small, and rather different from those modern forms just enumerated.

Sediment in typical context of the Burgess Shale deposit of Mid Cambrian age. The animals lived offshore, beyond a major reef-like structure, and they were apparently buried rapidly by the dramatic slipping of large amounts of sediment.

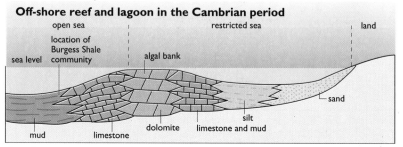

Off-shore reef and lagoon in the Cambrian period

Another major group of animals, the echinoderms, achieved considerable importance in the early Palaeozoic seas. Echinoderms include sea urchins, sea cucumbers, starfish, and crinoids, or sea lilies. The crinoids were abundant, most of them attached to rocks or hardened mud on the sea bed. They consist of a long stalk made from rings of calcite, and a cup-like structure at the top, which contains the body. The animal feeds by drawing particles of food from the water down the insides of its numerous tentacles, and into its mouth in the centre of the cup. Early sea urchins and starfish also patrolled the reefs in search of food: these were some of the first predators.

Many small animals, the plankton, floated in the top few metres of the sea. Some of the most unusual animals of the Palaeozoic plankton were the graptolites ('written stones', often preserved looking like pencil marks in the rock) which were especially abundant in the Ordovician and Silurian (440–410 million years ago), but are known through most of the Palaeozoic. They are used extensively for dating marine rocks of Ordovician and Silurian age.

Graptolites generally formed colonies, often of hundreds of individuals, and the colonies were contained in tough flexible tube-like structures that sometimes branched and coiled in complex ways. Each little creature sat in a cup-like structure, and these were arranged like the teeth on a saw blade. The animals fed on microscopic floating organic matter, which they absorbed through tentacles. It is thought that the graptolites kept afloat by being very lightweight, and in some cases by the use of gas-filled flotation structures like small balloons, and by the hydrodynamic properties of their skeletons. The complex spirals and branching patterns meant that the colonies sank slowly, and they rotated in some cases.

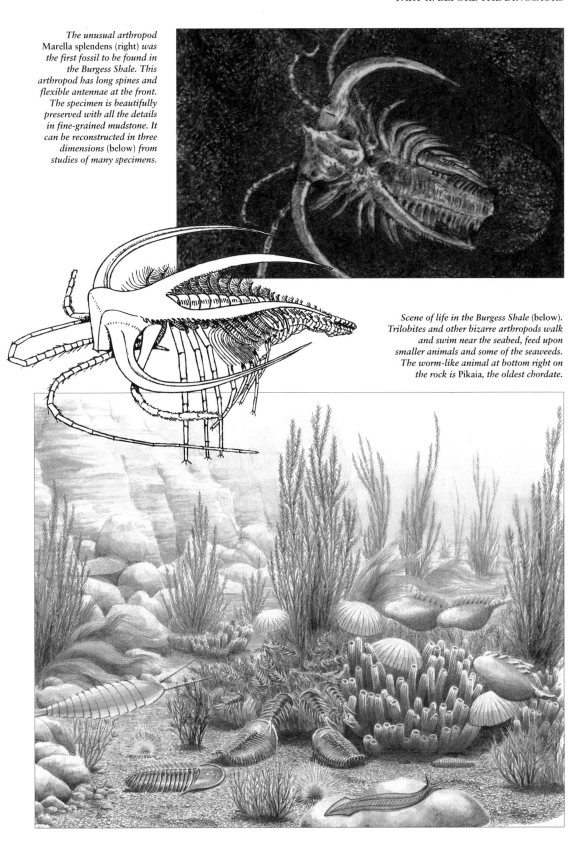

The unusual arthropod Marella splendens (right) was the first fossil to be found in the Burgess Shale. This arthropod has long spines and flexible antennae at the front. The specimen is beautifully preserved with all the details in fine-grained mudstone. It can be reconstructed in three dimensions (below) from studies of many specimens.

Scene of life in the Burgess Shale (below). Trilobites and other bizarre arthropods walk and swim near the seabed, feed upon smaller animals and some of the seaweeds. The worm-like animal at bottom right on the rock is Pikaia, the oldest chordate.

"There is no more ineresting story in the fossil record than that of the rise and fall of the reptiles. They became the dominating group of vertebrates during the whole of the Mesozoic and gave rise to many curious and spectacular types. Now they are in decay; but it must not be forgotten that the birds and mammals which have triumphed over them are their descendants."

Alfred Sherwood Romer, 1966

Animals with Backbones

The earliest remains of vertebrates (backboned animals) in Upper Cambrian sediments of North America, are the bony scales of *Anatolepis*, which are known also in the Lower Ordovician of North America, Greenland, and Spitsbergen. These were jawless animals with a lightweight internal skeleton, and a strong outer skeleton made from a mixture of bony tissues composed of the same materials as our teeth (dentine and enamel). Modern sharks have toothy denticles set into their skins. The outer skeletons of these earliest fishes, the agnathans ('no-jaws') were hard all-encompassing cases that enclosed the whole head area and front part of the body. Only the tail projected, as a flexible unit behind the shell, so that it could beat gently from side to side to drive the fish through the water.

There were two main agnathan groups in the Cambrian and Ordovician, the conodont ('cone-tooth') animals and the ostracoderms ('shell-skinned'). Conodonts were eel-shaped animals, 4–20 cm long, which did not have a body covering. At the head end were a pair of huge eyes, and a complex basket of teeth, consisting of thirteen separate elements that articulated together to make a flexible grasping and tearing apparatus. Until 1983, conodonts were known only from their tooth-bearing elements. The affinities of conodonts had been a complete mystery until an extraordinary discovery near Edinburgh, Scotland settled the question. This was a conodont apparatus in place at the front end of the body of a slender soft-bodied animal, preserved as a compressed filament on the rock. Since 1983, more conodont animals have come to light in Scotland and in South Africa.

The oldest complete ostracoderms are Ordovician in age, and they are known from many parts of the world. The group radiated in the Silurian, and especially in the Devonian (410–360 million years ago), when many groups are known from the Old Red Sandstone continent of northern Europe and North America, as well as China and Australia. The Devonian ostracoderms include small blunt-headed forms, such as *Hemicyclaspis*, that fed on debris in the mud, bullet-shaped swimmers, and some from China with an astonishing array of spines and crests on their heads. The mouth was a broad opening beneath the head shield. The bullet-shaped ostracoderms, such as *Pteraspis* from the Devonian of Scotland, were probably reasonably fast swimmers. *Pteraspis* has a moderately pointed snout, a round head shield shaped like the front end of a torpedo, and a flexible body and tail region that would have been capable of active side-to-side beating in order to power the animal through the water. These groups of jawless fishes did not outlive Devonian times, but the conodont animals lived on until the end of the Triassic. There are only two groups of living agnathans, the lampreys and hagfishes, eel-shaped fishes which feed largely parasitically on bony fishes by rasping flesh from their sides.

Fishes with Jaws

The first jawed vertebrates were acanthodians ('spiny'), a group that arose in the Mid Silurian, and survived until the Mid Permian, although they had their heyday in the Devonian. The acanthodians were small active-swimming fishes that flitted about in large shoals in the surface waters. Their small skeletons, less than 20cm long, are often found in vast numbers in rocks deposited in shallow seas and lakes. Acanthodians, such as

Euthacanthus, had stout spines at the front of each fin, and a further battery of paired spines along their bellies. Acanthodians probably required this astonishing armament simply to make them unpalatable: to swallow an acanthodian would be like swallowing an open Swiss army knife!

The origin of jaws in acanthodians, and in all later vertebrates, is a mystery. There is no complete series of fossils to show what happened, but it seems that jaws may be modified gill supports. Fishes have gill slits, through which water passes and oxygen is extracted. These fleshy gills are supported by thin rods of bone or cartilage, which are arranged in hinged pairs, each with an upper and a lower element. Perhaps the front gill support became attached to the skull, and the upper and lower hinged parts of these became the upper and lower jaws. The gill supports may support teeth, there is not a problem in imagining the migration of teeth on to the early jaws.

A second group of fishes with jaws were the placoderms ('plated skin'), known solely from the Devonian. A typical placoderm, such as *Bothriolepis*, was modest in size, and had a superficial resemblance to an ostracoderm. The head and shoulder region was enclosed in a bony box, and the back of the body was more flexible and covered with thick scales. However, the body armour of placoderms is quite different from that of the contemporary agnathans. The pattern of plates is different, and the head shield was also divided into mobile segments. The lower jaw was composed of a couple of plates that hinged at the back, and allowed the mouth to open and shut. The placoderms did not have teeth, but their place was taken by the sharp edges of the jaws: great triangular projections of the jaw bone like the point of a can opener. Placoderms may have fed close to the bottom of the sea, or lake, in which they lived, and they may have used their spike-like fins to stilt-walk over the mud.

Early fishes included Xenacanthus *and* Dunkleosteus *(below).* Xenacanthus *was a swift moving shark, with a spine behind its head.* Dunkleosteus *was a giant placoderm, a fish with a plated head, and no teeth. It cut its prey with sharpened bony plates.*

Sharks and Bony Fishes

Other Devonian fishes were more modern in appearance. The first chondrichthyans, or cartilaginous fishes, came on the scene in the late Devonian. Early forms, such as *Cladoselache*, were very shark-like in appearance. This fish was 2m long and sleek in outline, with a small head, a streamlined body, and broad pectoral (front) fins used for steering. Shark evolution only took off during the following Carboniferous period (360–290 million years ago), and some of these early sharks were most unusual. The stethacanthids, like *Falcatus*, had a bony spine above the shoulder region that bent forward as a broad tooth-covered projection. This may have been used in fighting or threat-displays prior to mating. Other stethacanthids, such as *Stethacanthus* itself, had a bony spine over the shoulder region which was shaped more like a shaving brush. On top of the structure was a flattened area which bore numerous sharp teeth. There was a further grouping of teeth on the forehead. The function of these teeth is utterly mysterious.

These sharks belong to primitive groups, and modern-style sharks appeared only much later, during the Mesozoic. Other Carboniferous chondrichthyans include chimaeras, or holocephalans, relatives of the modern catfishes, which have a large head, and slender pointed body ending in a long whip-like tail. Some of the Carboniferous holocephalans were longer and more slender than modern forms. Some had small heads and long tapering bodies while others had large heads, long front fins, and rounded tail fins.

The final fish group to appear in the Devonian were the osteichthyans, or bony fishes. There are two groups, those with ray-like fins, ancestors of most fishes today (more than 90% of species) from carp to salmon, and sea-horse to tuna, and the lobefins, which had thick muscular limb-like fins. The ray-fins of the Devonian include *Cheirolepis*, a fast-swimming predator which must have used its large eyes for locating prey. The advanced jaw apparatus with its great gape, allowed *Cheirolepis* to swallow prey up to two-thirds of its own size. The commonest food for *Cheirolepis* must have been the acanthodians, but this bony fish may also have hunted smaller lobefinned fishes and some of the less armoured agnathans and placoderms.

The Devonian lobefins are represented by some diverse groups, such as the rhipidistians and lungfishes. The rhipidistian *Eusthenopteron* was also long and slender, and was an active predator. These lobefins had muscular front fins, and could have used these to haul themselves over mud from pond to pond. *Dipterus*, a well known Devonian lungfish, reached a length of 20 cm, and it had a slender body with narrow fins fore and aft. Like its relatives, *Dipterus* did not have teeth round the margins of its jaws. Its dental battery consisted of broad grooved plates of dentine in the roof of the mouth, and some smaller tooth-like structures in front. These were used for grinding and crushing tough food of various kinds, perhaps arthropods and shelled invertebrates, perhaps some of the smaller armoured ostracoderm and placoderm fishes.

The Devonian bony fishes, Cheirolepis (top), Osteolepis (middle) and Dipterus (bottom), belong to three key groups, the ray fins, rhipidistians and lung fishes.

After the Devonian, the rhipidistians disappeared, and the coelacanths and lungfishes continued to evolve at a slow pace. Indeed, these two groups are often called 'living fossils', because the modern forms are very like their Devonian ancestors, but also because both groups remained rare and they changed only slowly. The three species of modern lungfishes, found in Australia, Africa, and South America, breathe air, and they are adapted to survive through dry periods by burying themselves in a cocoon in the mud, where they wait until the rains come again, and flood their ponds. The coelacanths are even more remarkable. It was thought that they had died out over 100 million years ago, until a living example was found. In 1938, a coelacanth, later named *Latimeria*, was fished out of deep waters off East Africa, and more have been caught since. This large fish swims and stilt-walks over the seabed. Its swimming style is extraordinary, consisting of walking-like movements of the various paired fins, with some wriggling of the body, but no side to side beating of the tail, as in most other fishes.

On to the Land

The first land plants were probably green algae that spread out of the water, perhaps as early as the Precambrian, and formed a green slime over damp pond sides. Later, perhaps mosses and liverworts evolved, growing in damp nooks and crannies on rocks. None of these simple plants, however, could venture far from the water, and they were all small. The first true land plants, the so-called vascular plants, are known from Silurian rocks. Vascular plants are characterised by the possession of tracheids, vascular conducting systems. Water and nutrients pass through the tracheids, and some structure of this sort was essential for land plants to become larger than mosses.

The oldest vascular plant is *Cooksonia*, a small plant composed of round stems that branch in two at various points and which are capped by spherical spore-bearing structures at the tip of each branch. *Cooksonia* is probably a member of the Rhyniopsida, a basal group that is known most fully from the Early Devonian Rhynie Chert of north-east Scotland which has exquisitely preserved numerous plants and arthropods in silica. The rhyniopsids from Rhynie had the same branching stems and terminal spore cases as Cooksonia, and they reached heights of 180mm. Groups of vertical stems were supported on horizontal branching structures which probably remained underwater or in damp areas close to the margins of ponds. The rhyniopsids and their relatives dwindled by mid-Devonian times, but their relatives, the horsetails, club mosses, and ferns became larger, and some of them achieved tree-like proportions during the Carboniferous and Permian (see pages 52–53).

The oldest true land plant, Cooksonia (above) was a small simple branching structure, first known from the Silurian of Wales.

Arthropods may have been the first animals to conquer the land. The first scorpions and spiders are known from the Late Silurian, in association with the oldest vascular land plants, and by the Devonian, several lines of insects, spiders, and scorpions had become established. By the Carboniferous, rich faunas of these, as well as millipedes and centipedes, burrowed through the leaf litter of the great coal forests in search of nourishment. Some of the Carboniferous terrestrial arthropods were truly spectacular. There was a 2m-long millipede, which left its broad tracks in several locations in Europe and North America. There was also a primitive dragonfly, *Meganeura*, which had a wingspan of up to 80cm.

Tetrapods, the four-legged land vertebrates, arose from lobefinned fishes during the Devonian. There has been some debate about the closest fishy relatives of the tetrapods, but most attention has focused on the rhipidistians, because of their complex bony and muscular pectoral (front) and pelvic (back) paired fins. It seems likely that these pairs of fins, lying respectively below the shoulder and hip region, correspond to the front and back legs of a tetrapod. It is possible to draw comparisons between the bones of the pectoral fin of a rhipidistian, such as *Eusthenopteron*, and those of the forelimb of an early amphibian. This search for equivalents between the bones of fish and tetrapod led to the classic view that the basic tetrapod limb bore five digits (fingers or toes), but recent work shows this is not the case.

An early amphibian (below) lived close to the water, but ventured on to land to hunt for worms and insects in the warm tropical forest.

The oldest tetrapods are Late Devonian in age. *Ichthyostega* from Greenland, a 1m long animal with four limbs, shows a perfect mix of tetrapod and fishy features. Land-living features include the strengthened rib cage and the clear separation between the head and shoulder girdle: these are joined in the rhipidistians, but had to separate when animals started to walk, or the vibrations in the shoulder bones would have shaken the early tetrapod's ear region unmercifully. *Ichthyostega* shows its rhipidistian heritage, however, by the fin on its tail, and by the arrangement of the bones of its skull which are almost identical in detail to those of *Eusthenopteron*.

The story of the fingers and toes turns out to be more complicated than had been thought. New studies, in the 1990s, of *Ichthyostega* and *Acanthostega*, both from the Late

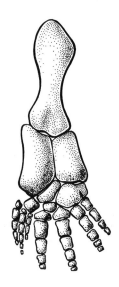

The leg of the early amphibian Ichthyostega, showing that it had seven toes. This discovery, in the 1990s, came as a surprise, since palaeontologists had always assumed that five fingers and toes were the fixed number.

Devonian of Greenland, and of *Tulerpeton* from Russia, show that the first tetrapods had more than five fingers and toes, indeed as many as seven or eight. This caused a major rethink of the classic story of the evolution of vertebrate limbs: five digits must have become standard only after the origin of tetrapods. Some time after their origin, tetrapods clearly settled on the standard five digits, which we still have. So much for the decimal system of counting!

In moving from a life in water to life on land, the first tetrapods faced major problems. The main problem was support: in water, an animal 'weighs' virtually nothing, but on land the body has to be held up from the ground, and the internal organs have to be supported in some way within a strong rib cage to prevent them from collapsing. In fishes, the backbone is simply an attachment point for various muscles used in swimming, but in tetrapods the backbone also supports the weight of the body which hangs between the front and hind legs. So the backbone has to be strengthened to act like the spanning portion of a bridge, and this involves strengthening of the bones, firmer attachments to the limb girdles, and new muscles and ligaments to tie it all together. In tetrapods, the hip bones are firmly fused to the backbone, as they are in humans, and the shoulder girdle is firmly attached to the rib cage. In addition, new muscles, and stronger neck vertebrae, are needed to hold the head up, not a problem faced by fishes. The limbs and limb muscles also become modified to allow walking, a rather jerky way of moving when compared with fluid grace of swimming.

Of course, most other body systems had to become modified for life on land. The lungs took over more and more the function of breathing. In addition, reproductive, osmotic (water balance), and sensory systems had to adapt, but here the changes did not happen all at once. The first tetrapods probably still spent a great deal of time in the water, and they continued to produce aquatic young (tadpoles) just as fishes do. Water balance was not a problem so long as these early tetrapods remained damp. Sensory systems used in the water, such as the electrically sensing lateral line, were retained in the first tetrapods. Only gradually did sight, smell, and hearing take on more important roles for the early tetrapods in sensing their environments. Amphibians are halfway to land life, but they retain many water-living adaptations.

The first amphibians include the temnospondyls, important in Carboniferous communities, and they continued with reasonable success through the Permian and Triassic, finally dying out in the Early Cretaceous. Temnospondyls had a low round-snouted skull , and most of them appear to have operated like sluggish crocodiles, living in or near fresh waters, and feeding on fishes. Some temnospondyls became fully terrestrial, and others evolved elongate gavial-like snouts for catching rapidly-swimming fishes. Temnospondyls had tadpole young, just as modern amphibians, such as frogs and salamanders, do. The tadpoles are occasionally found as fossils, typically in conditions of exceptional preservation, where the tiny bodies are preserved in black muds. Two groups of related smaller Carboniferous and Permian amphibians were the nectrideans and microsaurs, relatives of the temnospondyls. The nectrideans were all aquatic animals. A second Carboniferous amphibian lineage led to the reptiles. Their main representatives in the Carboniferous and Permian were the anthracosaurs, which had longer narrower skulls than the temnospondyls, but may have had similar lifestyles, hunting prey on land and in fresh waters.

Origin of the Reptiles

The oldest-known reptile, *Hylonomus* from the Mid Carboniferous of Canada, has been superbly well preserved inside ancient tree stumps, into which it crawled in pursuit of insects and worms, and then became trapped. *Hylonomus* looks little different from some amphibians of the time, such as the microsaurs, but it shows several clearly reptilian characteristics, a high skull, evidence for additional jaw muscles, and an astragalus bone in the ankle. One key reptilian characteristic, and indeed a characteristic of all reptiles and their descendants, the birds and mammals, is not known in *Hylonomus.* This is the cleidoic egg, the key to the success of these essentially fully terrestrial groups.

Reptiles, unlike amphibians, have broken with aquatic reproduction by enclosing their eggs within a tough semi-permeable shell, hence the term cleidoic ('closed'). The shell is usually hard and made from calcite, but some lizards and snakes have leathery egg shells. The shell retains water, preventing evaporation, but allows the passage of gases, oxygen in, and carbon dioxide out (see picture). The developing embryo is protected from the outside world, and there is no need to lay the eggs in water, nor is there a larval stage in development. Inside the egg shell is a set of membranes that enclose the embryo (the amnion), that collect waste (the allantois), and that line the egg shell (the chorion). Food is in the form of yolk.

Reptiles radiated during the Late Carboniferous, and the three main lines became established. They are distinguished by the pattern of openings in the side of the skull. The primitive state is termed the anapsid ('no hole') skull pattern, since there are no temporal openings. The two other skull patterns seen in amniotes are the synapsid ('same hole'), where there is a lower temporal opening, and the diapsid ('two hole') pattern, where there are two temporal openings. These temporal openings correspond to low-stress areas of the skull, and the edges serve as attachment sites for jaw muscles.

These three skull patterns define the key reptile groups. The Anapsida include various early forms such as Hylonomus, as well as some Permian and Triassic reptiles, and the turtles. The Synapsida include the mammal-like reptiles (see pages 54–55) and their descendants, the mammals, and the Diapsida includes a number of early groups, as well as the lizards and snakes, and the crocodiles, pterosaurs, dinosaurs, and birds. The three great reptile groups had great success during the Permian, but they were nearly wiped out during an astonishingly severe mass extinction 250 million years ago (see pages 56–57). This extinction event, however, may be said to have cleared the way for the later evolution of the dinosaurs.

Chorion Embryo

lantois
aste material)

Amnion

Yolk Sac
(food)

shell

The cleidoic egg of reptiles, birds, and primitive mammals (above) has a hard calcareous shell, and several internal membranes that control the passage of water and oxygen.

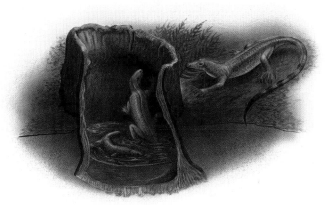

The oldest reptile, Hylonomus, *from the Mid Carboniferous of Nova Scotia (right).* Hylonomus *has a high skull, a reptilian feature, and it almost certainly laid eggs with hard shells. Its remains are preserved in ancient hollow tree stumps which acted as pitfall traps.*

Timeline I

PRECAMBRIAN	PALEOZOIC

PRECAMBRIAN

4,000 million years ago

Archean
3,300 million years ago
Oldest sedimentary rocks.
First stromatolites.
Atmosphere with some carbon dioxide.

2,900 million years ago
Massive stromatolites formed by photosynthesizing blue-green algae.

2,500 million years ago

Proterozoic
1,800 million years ago
Diversification of species of prokaryote algae (cellular forms with no nucleus).

1,200 million years ago
Development of eukaryote cells. These cells have a nucleus containing DNA, and the capacity for sexual reproduction.

600 million years ago
Appearance of diverse species of soft-bodied, multicellular organisms (Ediacaran Fauna).

550 milion years ago

PALEOZOIC

550 million years ago

Cambrian
550 million years ago
Worldwide emergence of marine invertebrate groups with shells and skeletons. Trilobites, brachiopods, archeocyathids, echinoderms, molluscs all common. Stromatolites decline in abundance.

500 million years ago

Ordovician
480 million years ago
First definite vertebrates – jawless freshwater fish. Freshwater plants assumed to be present.

450 million years ago
Possible first land plants.

440 million years ago

Silurian
440 million years ago
Abundance of jawless fish. First fish with jaws – freshwater acanthodians. Giant sea-scorpions (eurypterids) emerge.

420 million years ago
First land plants. Vascular plants including lycopsids and psilopsids present, but very rare. First insects and arachnids.

410 million years ago

Devonian
400 million years ago
Age of fishes. Jawed and armoured fish become abundant and diversify. Development of modern types of fish with bony skeletons and scales. Some fish groups develop lungs. Spore-bearing plants become more common on land – though still tied to aquatic habitats.

370 million years ago
The first amphibians develop from fish and reach the land. Emergence of sea ferns, while true ferns cover some lowland areas in dense forest.

350 million years ago

Carboniferous
340 million years ago
First true reptiles. Emergence of distinct floras associated with different climatic conditions. *Glossopteris* flora dominates Gondwanaland.

300 million years ago
Development of huge lycopsid plants in swamp forests. Amphibians and reptiles diversify in humid tropical conditions, as do insects. Abundance of giant flying insects and cockroaches.

290 million years ago

Permian
270 million years ago
As conditions become drier and hotter, reptiles thrive at the expense of amphibians. Development of warm-blooded reptiles (therapsids) the precursors of the mammals.

250 million years ago
Mass extinction of marine life. Groups made extinct include trilobites, rugose corals and crinoids. Other marine invertebrates severely affected. Fish are generally unaffected.

350 million years ago

250 million years ago

MESOZOIC

250 million years ago

Triassic
250 million years ago
Ammonites survive the mass extinction at the end of the Paleozoic and thrive in the Mesozoic. Development of thecodontian reptiles.

230 million years ago
Dinosaurs evolve from thecodont reptiles. First mammals emerge from warm-blooded therapsid reptiles.

205 million years ago

Jurassic
205 – 145 million years ago
Dinosaurs become dominant, reaching their largest size. Diversification of flying reptiles (pterosaurs) and aquatic reptiles (plesiosaurs). Continued diversification of insects. *Archaeopteryx*, the earliest known bird (or feathered dinosaur), evolve.

145 million years ago

Cretaceous
145 – 65 million years ago
Continuing dominance of land by dinosaurs. Mammals remain small. Reptiles diversify – turtles, snakes, lizards are abundant. Emergence of flowering plants (angiosperms). These dominate the land plant kingdom by the end of the Cretaceous.

65 million years ago
Mass extinction of marine and land life-forms. Principal casualties are the dinosaurs and ammonites.

65 million years ago

TERTIARY

65 million years ago

Paleocene
65 million years ago
Reptile groups (other than dinosaurs) survive the mass extinction. Mammals and birds also survive and flourish. Emergence of early horse, elephant and bear groups of mammals. Compositae family of plants emerge.

53 million years ago

Eocene
50 million years ago
Grasses emerge and diversify rapidly along with Leguminosae and Compositae plants.

40 million years ago
Grazing animals and monkeys emerge. Mammal groups (whales, dolphins) return to the sea. Foraminifera grow and diversify.

36 million years ago

Oligocene
35 million years ago
The first apes emerge. Large mammals and birds spread over the Earth. Grasses cover large areas of land.

23 million years ago

Miocene
10 – 11 million years ago
Separation of great apes and hominid apes. Radiation of hominid primates culminates in *Sivapithecus* – an ape showing many characteristics of living apes and humans.

5 million years ago

Pliocene
3 – 4 million years ago
Emergence of *Australopithecus*. First hominids.

2 million years ago

QUATERNARY

2 million years ago

Pleistocene
2 – 1.75 million years ago
Homo habilis, and a possible early form of *Homo erectus*, emerges in East Africa.

1 million years ago
Homo erectus disperses from Africa as far as China and Java.

500,000 years ago
Homo sapiens (Neanderthal man) appears in Africa and migrates to Europe.

90,000 years ago
Modern humans (*Homo sapiens sapiens*) emerge in southern Africa.

50,000 years ago
Modern humans reach the Middle East.

35,000 years ago
Modern humans reach Europe as Cro-Magnon man.

20,000 – 10,000 years ago
Modern humans enter North America via the Bering land bridge and move south.

10,000 years ago

Holocene
10,000 years ago
Homo sapiens sapiens reaches every continent except Antarctica, and is the only surviving hominid.

present day

Continents on the Move

Over millions of years the continents have moved their positions, and this process has affected the evolution of life.

There is a vast array of evidence indicating that the continents are not stable. In the 19th century, geographers noticed the apparent match in shape of the eastern coast of South America and the western coast of Africa, and concluded that the two continents may once have fitted together and have since drifted apart. This idea was taken forward early in the 20th century, when Alfred Wegener proposed a full-blown model of continental drift. His evidence was not only the relative shapes of the continents, but also a number of unexplained observations about ancient rocks and fossils.

Wegener gathered some impressive geological evidence for continental drift. He noticed that certain major rock formations which extend to the east coast of Brazil are matched by identical formations in west Africa. When modern geological maps of South America and Africa are pushed together, many features match, as if a great jigsaw has been put back together. He also observed that the distribution of many fossil groups makes no sense when studied on modern maps. The Permian tree *Glossopteris*, for example, is known from South America, southern Africa, India and Australia, in a distribution extending from near the present-day South Pole to India, now north of the Equator. If those southern continents, together with India and Antarctica, were pushed back together to form a supercontinent (Gondwanaland), then the distribution of *Glossopteris* would make sense.

When viewed from the South Pole, it can be seen that the east coast of South America and the west coast of Africa could fit together. Antarctica swings in towards the east coast of Africa, and the curve of the south coast of Australia fits neatly against the western margin of Antarctica. Two hundred million years ago these continents, together with India, were united as the supercontinent known as Gondwanaland.

Many geologists opposed Wegener's theory of continental drift, mainly because they were unable to imagine a process that could drive great masses of the Earth's crust about. It was only in the 1950s and 1960s that this process was discovered. Great convection currents of molten magma occur deep within the Earth, in a zone called the mantle. These currents rise, then turn sideways below the solid crust. This crust is divided into a number of large and small plates, some lying below the oceans, others forming parts of the continents. Slowly, at rates of a few centimetres per century, the rising convection currents move these plates. In places they move apart, and a new crust is created from the molten magma. In other places, one plate dives beneath another, and crustal rocks are consumed in the molten mantle below.

1

During Cambrian times, 500 million years ago, the continents were arrayed along the equatorial belt. Only Gondwanaland extended into temperate latitudes.

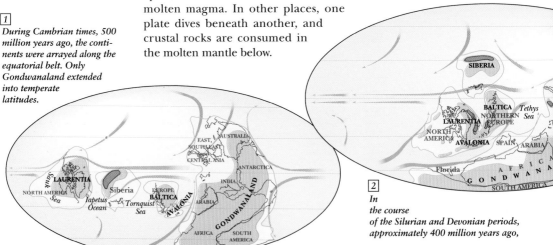

2

In the course of the Silurian and Devonian periods, approximately 400 million years ago,

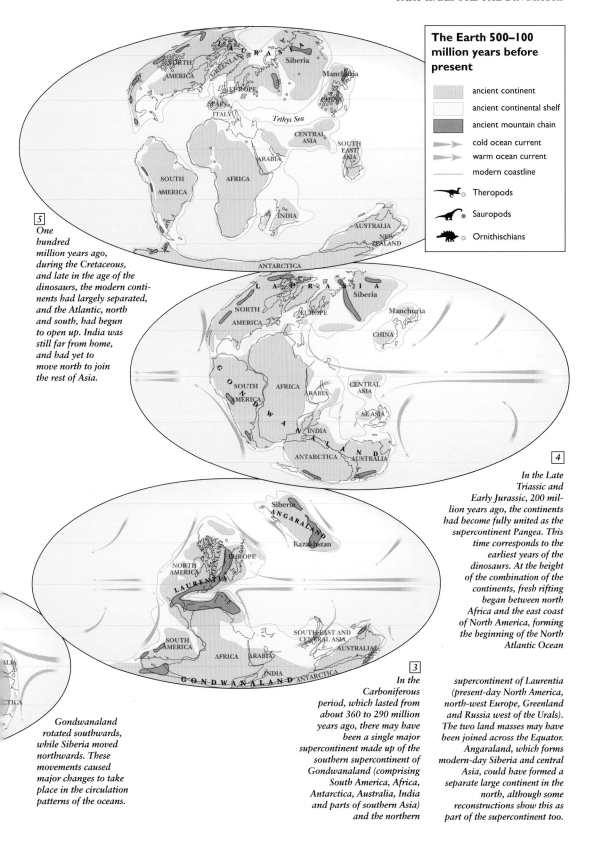

The Earth 500–100 million years before present

	ancient continent
	ancient continental shelf
	ancient mountain chain
	cold ocean current
	warm ocean current
	modern coastline
	Theropods
	Sauropods
	Ornithischians

5 *One hundred million years ago, during the Cretaceous, and late in the age of the dinosaurs, the modern continents had largely separated, and the Atlantic, north and south, had begun to open up. India was still far from home, and had yet to move north to join the rest of Asia.*

4 *In the Late Triassic and Early Jurassic, 200 million years ago, the continents had become fully united as the supercontinent Pangea. This time corresponds to the earliest years of the dinosaurs. At the height of the combination of the continents, fresh rifting began between north Africa and the east coast of North America, forming the beginning of the North Atlantic Ocean*

3 *In the Carboniferous period, which lasted from about 360 to 290 million years ago, there may have been a single major supercontinent made up of the southern supercontinent of Gondwanaland (comprising South America, Africa, Antarctica, Australia, India and parts of southern Asia) and the northern supercontinent of Laurentia (present-day North America, north-west Europe, Greenland and Russia west of the Urals). The two land masses may have been joined across the Equator. Angaraland, which forms modern-day Siberia and central Asia, could have formed a separate large continent in the north, although some reconstructions show this as part of the supercontinent too.*

Gondwanaland rotated southwards, while Siberia moved northwards. These movements caused major changes to take place in the circulation patterns of the oceans.

Carboniferous and Permian World

The origin of reptiles and their rise to dominance.

One of the oldest reptiles, Hylonomus from the Mid Carboniferous of Nova Scotia, Canada. This was a small insect-eating animal with a high skull and strong jaw muscles. It almost certainly laid eggs with shells, and this allowed it to hunt its prey well away from the water.

The Carboniferous and Permian periods were marked by major changes, which were generally warm in the northern hemisphere. A large ice cap developed in the southern hemisphere, and covered large parts of Gondwanaland during the Late Carboniferous and the Early Permian. There is good evidence for the ice cap in the form of glacial striations, deep grooves gouged into rocks by the passage of glaciers which dragged great boulders along at their base. These are seen in South Africa, South America, India, and Australia, indicating a centre of glaciation in southern Africa. Sea levels were also lowered worldwide, as large volumes of water from the oceans were locked up in the great southern ice cap. During the Early Permian, the ice melted, and sea levels rose.

Plants changed dramatically in the Carboniferous. Before that time, most plants had been relatively small, and they had been largely restricted to the damp areas around ponds. In the Carboniferous, however, many large and exotic-looking trees evolved, giant horsetails, seed ferns, and the like, some of them as much as 20 or 30 metres high. These trees formed lush tropical forests over much of Europe and North America, and indeed, they are the basis of much of the commercialy exploited coal in those areas. Among the roots of the trees lived a multitude of insects, spiders, and millipedes, and giant insects flapped between the branches. Fishes swam in the ponds and coastal seas.

During the Carboniferous and Permian, the continents moved closer and closer, until there was essentially a single super-continent, Pangea, at the end of the Permian. A great ice sheet covered the southern continents during the Late Carboniferous and Early Permian.

In the Early Carboniferous, the amphibians gave rise to the reptiles. The first reptiles were small lizard-sized animals, but they showed major changes from their amphibian ancestors. The first reptiles had waterproof skins, and they laid eggs with thick shells. This meant that it was unnecessary for them to remain near water to keep moist, or to lay their eggs in water, as amphibians do.

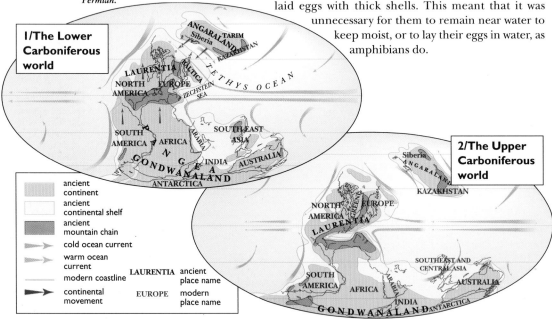

1/The Lower Carboniferous world

2/The Upper Carboniferous world

ancient continent	
ancient continental shelf	
ancient mountain chain	
cold ocean current	
warm ocean current	
modern coastline	LAURENTIA ancient place name
continental movement	EUROPE modern place name

3/The Permian world

A scene in the Early Permian of Germany, showing newt-like and crocodile-like amphibians hunting fishes in a shallow pond. Some small amphibians also move in and out of the water, and the sail-backed reptiles, Edaphosaurus *and* Dimetrodon, *take the air at the top, as they seek food.*

The Land that Time Forgot

The reptiles of the Karoo Basin in South Africa tell a story of dramatic success.

The central region of South Africa is a great desert-like bowl called the Karoo Basin. The scrub-covered land is mainly rock, and it offers poor returns to the farmers. However, this great red rocky land has been a dramatic source of specimens of mammal-like reptiles of Late Permian age, the synapsids. The synapsids arose in the Late Carboniferous, and during the subsequent Triassic period, one group of synapsids, the cynodonts, became more and more mammal-like, until they gave rise to the first true mammals in the Late Triassic (see pages 59–60).

The synapsids of the Karoo were found first in the 1850s by Alexander Bain, a Scottish engineer. He sent them back to England, where Sir Richard Owen (see pages 22–23) described them. British colonists continued collecting, and they found the Karoo rocks were littered with skeletons. Soon, the collectors tired of extracting all the bones, and they concentrated simply on picking up skulls and returning them to museums in England, and eventually to museums in Cape Town and Johannesburg. Hundreds of species of synapsids were named between 1920 and 1950 by Robert Broom, another Scot, this time a physician, who had settled in South Africa.

The Karoo synapsids include meat-eating and plant-eating dinocephalians, some of them quite large. *Moschops*, in particular, was an extraordinary animal, equipped with massive shoulders and neck and tiny limbs and head. *Moschops* also had a massively thickened skull roof, and it has been suggested that it engaged in stentorian head-butting contests in seeking mates. The meat-eating dinocephalians were more lithe and superficially dog-like.

The meat-eating dinocephalian Titanosuchus *(below), a slender hunter that fed on small herbivorous reptiles, and probably also on defenceless juveniles. Young dinocephalians may also have subsisted on worms and insects. Some dinocephalians, like* Moschops *(opposite, top) were plant-eaters, and these thick-skinned reptiles were preyed on by sabre-toothed gorgonopsians like* Sauroctonus. *Smaller reptiles, like* Youngina, *fed on insects. These reptiles have been found in the Permo-Triassic rocks of the Karoo Basin (opposite, below), a huge area of ancient dry scrubland that supported a rich flora and fauna.*

The commonest herbivores were the dicynodonts, animals with either no teeth at all, or merely a pair of tusks. They sliced vegetation with sharp horn-lined beaks, and ground it with a circular jaw motion. Dicynodonts were bulbous animals, with large bodies, short legs, and inadequate tails. The dicynodonts and large plant-eating dinocephalians were preyed upon by the first sabre-tooths, the gorgonopsians. Gorgonopsians were powerful sleek animals with massive tusks. They could drop the lower jaw clear of the fangs, and then leap at a herbivore with the teeth clear and pierce their thick skins. The dinocephalians and gorgonopsians did not survive the end of the Permian, but the dicynodonts flourished a second time during the Triassic period.

ATLANTIC OCEAN

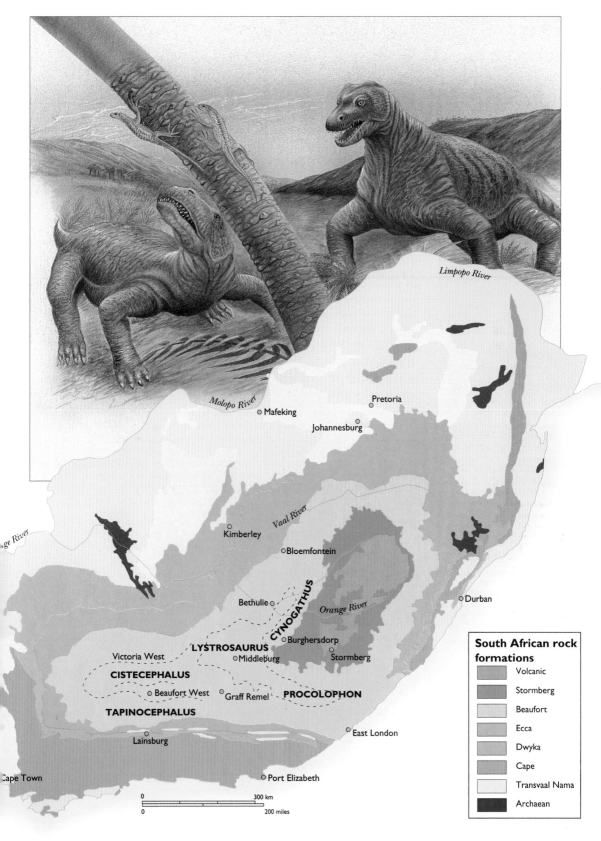

Limpopo River

Molopo River

○ Mafeking

Pretoria ◉

Johannesburg ○

ge River

Vaal River

Kimberley ○

○ Bloemfontein

○ Durban

Bethulie ○

Orange River

CYNOGATHUS

○ Burghersdorp

LYSTROSAURUS

Stormberg ○

Victoria West

○ Middleburg

CISTECEPHALUS

○ Beaufort West   ○ Graff Remel   PROCOLOPHON

TAPINOCEPHALUS

East London ○

○ Lainsburg

Cape Town

○ Port Elizabeth

| 0 | | | | 300 km |
| 0 | | 200 miles | | |

South African rock formations

	Volcanic
	Stormberg
	Beaufort
	Ecca
	Dwyka
	Cape
	Transvaal Nama
	Archaean

The End–Permian Mass Extinction

Two-hundred and fifty million years ago all life on earth came extraordinarily close to total extinction.

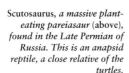

The biggest mass extinction of all time happened at the end of the Permian. It has been estimated that more than half the families that existed in the Late Permian disappeared, and that has been scaled up to imply that perhaps 95% of species died out. The reasoning behind this calculation is that each family of plants or animals normally contains dozens of species. If more than half the families died out completely, then most other families must have been very hard hit as well. So, a loss of 50% of families implies a loss of 95% of species.

In the seas, many groups of invertebrates died out: trilobites, rugose and tabulate corals, major groups of brachiopods, and many more. In fact, the seas immediately after the extinction event must have been empty and eerie places, with only a few surviving molluscs and shrimp-like creatures here and there. One or two sharks and bony fishes also survived, but numbers were sadly depleted. Coral reefs, and all the complexes of species that lived on and around the reef, disappeared completely, and when reefs finally recovered they were made up of quite different species.

On land too, the effects were devastating. The primitive amphibian groups (see pages 52–53) nearly disappeared completely, and most of the mammal-like reptiles (see pages 54–55), the dinocephalians, gorgonopsians, and many other groups, vanished. The record of faunal change on land is documented best in the Karroo Basin of South Africa (see pages 54–55) and in

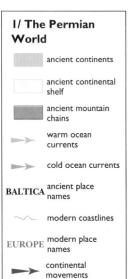

Scutosaurus, a massive plant-eating pareiasaur (above), found in the Late Permian of Russia. This is an anapsid reptile, a close relative of the turtles.

The Permian witnessed major fusions of the continents. Angaraland (modern Siberia) was the last major landmass to join.

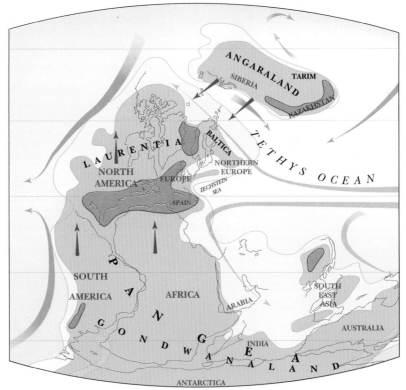

I/ The Permian World

- ancient continents
- ancient continental shelf
- ancient mountain chains
- warm ocean currents
- cold ocean currents
- **BALTICA** ancient place names
- modern coastlines
- EUROPE modern place names
- continental movements

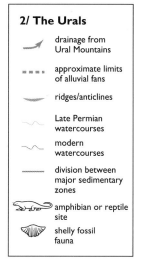

2/ The Urals

- ↗ drainage from Ural Mountains
- ╍╍╍ approximate limits of alluvial fans
- ～ ridges/anticlines
- ～ Late Permian watercourses
- ～ modern watercourses
- ── division between major sedimentary zones
- 🦎 amphibian or reptile site
- 🐚 shelly fossil fauna

The Ural Mountains were elevated by the fusion of Angaraland, modern Siberia, and the rest of Europe. During the Late Permian and Early Triassic, the lowland area to the west of the rising mountain chains was home to a diverse fauna of reptiles, many of them like those of the Karroo in South Africa (see pages 54–55). As the mountains rose higher, great river systems spread masses of sediment in large alluvial fans over the low-lying reptile habitats.

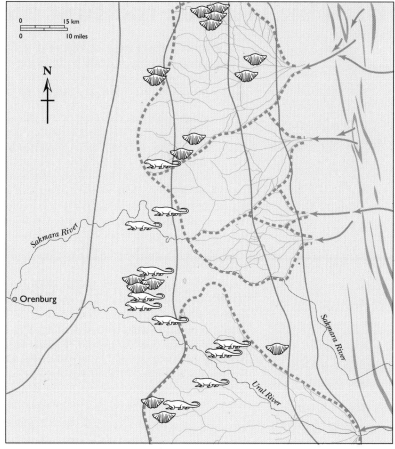

the Urals region of Russia. In both areas, the rich Late Permian faunas were decimated, and only three or four species survived.

The causes of this event are far from certain. Perhaps some aspects of continental drift (see pages 50–51) were responsible. As continents fused, rich areas of shallow sea life were wiped out, and the continents became more uniform, and formerly isolated species interacted, and perhaps many became extinct. This kind of model seems too long-term, however, to explain the dramatic extinctions.

Other changes were happening in the Late Permian. Some geologists argue that climates became colder, while others find evidence for dramatic heating. There were also huge changes in sea level, perhaps in part associated with the fusion of the continents, and the loss of inland seas.

The most recent ideas about the causes of the end-Permian extinction focus on the Siberian flood basalts. Over a time span of half a million years, precisely at the Permian-Triassic boundary, vast volumes of molten basalt lava were poured out of numerous volcanic vents in Siberia. During the eruptions, as much as three million cubic kilometres of lava were spewed out, and this covered an area the size of Europe. The lava and ash would have killed all life in eastern Russia, but there would also have been major climatic changes. Volcanoes release huge amounts of gas into the atmosphere, particularly carbon dioxide and sulphur dioxide, and these caused dramatic heating or cooling respectively, no-one knows which.

III: The Triassic: the first dinosaurs

The Triassic period marks the beginning of the age of the dinosaurs, but dinosaurs did not appear at the beginning of this time period. They were at first rare animals of modest size, and only after the extinction of competitor groups, did they radiate dramatically in the Late Triassic.

"We now come to a new phase in the history of vertebrate life, the development of back-boned animals during the Triassic period, the earliest subdivision of the Mesozoic era. The Triassic period was important as a time of transition between the old life of the Paleozoic era and the progressive and highly varied new life of the Mesozoic"

E.H. Colbert, 1969

Some of the key episodes in tetrapod evolution occurred during the Triassic period (250–205 Myr). This time marked the transition from faunas of Palaeozoic-style amphibians and reptiles, such as the mammal-like reptiles and pareiasaurs of the Late Permian (see pages 54–55) to more modern faunas. The unbelievable killing effects of the end-Permian extinction event (see pages 56–57), whatever the causes, had wiped out most animals and plants, and the new world of the Triassic must have been an eerie place, empty of life, and with only very restricted faunas and floras. It took some 10 million years or more before life had recovered to something like normal ecosystems, and it took longer for larger animals, for coral reefs, and for other specialised organisms to recover.

In many respects, the Triassic world was similar to that of Permian times. All continents remained united as the supercontinent Pangea (see pages 68–69), and climates were generally warm. Triassic life at first looked like a modified re-run of what had existed before in the Permian, but there were differences, and these became ever more marked in the Late Triassic. On land, the mammal-like reptiles re-radiated during the Triassic, but they had already lost a number of their key adaptive zones to two new groups—the archosaurs and the rhynchosaurs. In the seas, several lines of fish-eating reptiles emerged, the nothosaurs, placodonts, and ichthyosaurs.

The Late Triassic was a key episode in the evolution of tetrapods. Not only did the dinosaurs arrive on the scene, but a number of other major groups also arose: the crocodilians, the pterosaurs, the turtles, and the mammals.

Triassic Mammal-like Reptiles

Several groups of mammal-like reptiles survived into the Triassic. One group, the dicynodonts, had been important herbivores in the Late Permian (see pages 54–55). They dwindled, and virtually disappeared at the end of the Permian, but one genus, *Lystrosaurus*, survived, and radiated worldwide. These animals were found first in the Karroo Basin of South Africa, but they were later found in Antarctica, India, South America, China, and Russia, evidence for a global supercontinent at the time (see pages 50–51). *Lystrosaurus* was a modest-sized, plant-eating dicynodont, 1–2 metres in length, and heavily built. It was formerly interpreted as having been a semi-aquatic animal, possibly because its remains were found in water-laid rocks. New studies of its anatomy show that it had no special adaptations for swimming, and the skeletons are found in water-laid rocks since that is where the carcasses ended up. The fact that skeletons are found in river sandstones does not mean that *Lystrosaurus* actually lived in those rivers.

The first faunas of the earliest Triassic were dominated by the dicynodont *Lystrosaurus* to the extent that it represents more than 90% of all skeletons wherever it is found. The other associated animals include small meat-eating mammal-like reptiles, therocephalians and cynodonts, the early archosaur *Proterosuchus*, prolacertiforms, and procolophonids (see below). This extraordinary imbalance of the *Lystrosaurus* faunas is not a natural kind of ecosys-

The cynodont Thrinaxodon *(below), a reptile that shows mammal-like characters. It has lumbar ribs, which suggest it may have had a diaphragm, necessary to maintain breathing while running, and found in animals with high metabolic rate. The teeth are differentiated, and there is a zygomatic arch, or cheek bone, below the eye socket and temporal fenestra.*

tem. Later in the Triassic, dicynodonts became very large, up to 3 or 4 metres long, and weighing over a tonne. These animals were dominant herbivores in many parts of the world, but the group disappeared 225 million years ago in the Late Triassic, at the boundary between the Carnian and Norian stages.

The cynodonts were another important group of Triassic mammal-like reptiles. Cynodonts had arisen in the latest Permian, and in the Early Triassic, some like *Thrinaxodon*, looked very dog-like and probably had fur. In the snout area of the skull, there are numerous small canals which indicate the presence of nerves serving the roots of sensory whiskers. If *Thrinaxodon* had whiskers, it clearly also had hair on other parts of its body, and this means insulation and temperature control. This is strong evidence that the Triassic cynodonts were already warm-blooded, just as mammals are.

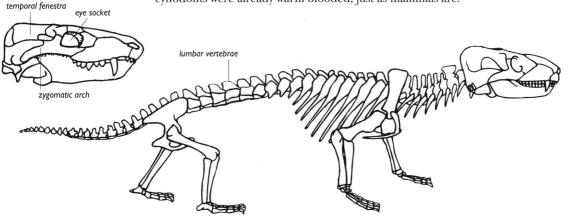

The Origin of the Mammals

There is much more evidence that mammals arose from cynodonts. For example, the teeth of *Thrinaxodon* are divided into three quite different shapes. The front teeth, or incisors, are short spike-like teeth, used to seize and nip at prey animals. Behind them is a single canine tooth on each side of the jaw. These long tusk-like teeth were used for killing prey animals by piercing their blood vessels. Behind the canines are the cheek teeth, which have blunter tops, and they were used for crushing bones and chewing flesh. Humans today, like all other mammals, share the same dental inheritance from the Triassic cynodonts, although our teeth are adapted for eating plants and flesh.

Cynodonts evolved along several lines during the Triassic, most of them small and medium-sized meat-eaters, but a few lines became modified to feed on plants. Some of the meat-eating cynodonts of the Mid and Late Triassic were extraordinarily mammal-like, and it is hard to draw a dividing line between reptiles and mammals in the middle of this astonishing sequence of 'missing links'.

The transition from mammal-like reptile to mammal is marked by a shift in the jaw joint, and an astonishing change of function of the bones at the back of the reptile jaw, which entered the middle ear of mammals and assist now in sound transmission. The typical reptile pattern is to have seven bones in the lower jaw, the dentary, which carries the teeth, the articular at the back which forms the jaw joint, and five others. The articular on the lower jaw meets the quadrate in the skull. In some advanced cynodonts there is a second jaw joint, between the dentary and the squamosal in the side of the skull wall. Mammals have only one bone in the lower jaw, the dentary, and the other jaw bones have changed function to operate in hearing.

Amphibians and reptiles have a single hearing bone, the stapes, which transmits vibrations from the ear drum on the side of the head to the inner ear. In mammals, there are three auditory ossicles, or hearing bones, the stapes, and the malleus (hammer) and incus (anvil). The last two are the articular and quadrate respectively, and these three tiny bones are delicately jointed. So the reptilian jaw joint is still there, hidden away in our middle ear, and the shift in function can still be detected when we hear our chewing.

The Diapsids Take Over

The diapsid reptiles, those with two jaw muscle openings in the back region of the skull (see page 47) were initially small to medium-sized carnivores that never became very abundant during the Carboniferous and Permian. However, things began to change during the Triassic, perhaps as a result of the end-Permian extinction event, which had such a devastating effect on therapsid communities. Modest-sized meat-eaters such as *Proterosuchus* appeared in the Early Triassic, and lived as a member of the *Lystrosaurus* fauna. This was one of the first archosaurs ('ruling reptiles'), a group that was later to include the dinosaurs, pterosaurs, crocodiles, and birds. Archosaurs have an additional skull opening between the eye socket and the nostril, termed the antorbital fenestra, the function of which is unclear.

Proterosuchus and other Early Triassic archosaurs took over the carnivorous niches formerly occupied by the gorgonopsids and titanosuchids that had died out at the end of the Permian. *Proterosuchus* was a slender 1.5m long animal that preyed on small and medium-sized mammal-like reptiles (therocephalians, dicynodonts) and procolophonids. It has short limbs that bend outwards at the knee, which shows that it had a sprawling posture, as in most Permian mammal-like reptiles and living lizards and salamanders. Further basal archosaurs of the Early Triassic include large predators up to 5 metres long, and some small possibly bipedal hunters.

The rhynchosaur Hyperodapedon, *from the Late Triassic of Scotland, has a deep skull and interlocking jaws lined with multiple rows of teeth.*

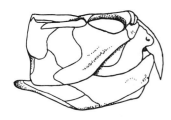

Other kinds of diapsids were important in the Triassic, the rhynchosaurs and prolacertiforms. These animals, widely different in external appearance, are close relatives of the archosaurs. Rhynchosaurs such as *Hyperodapedon*, from the early Late Triassic of Elgin, Scotland, had a curiously smiling face, with broad curved upper jaws, and a pair of tusk-like bones at the front. The lower jaw is deep, and it bears two rows of teeth, one on the crest, and the other lower down on the inside. The lower jaw clamped firmly into the groove on the upper jaw, just like the blade of a penknife closing into its handle. This firm closing action of the jaw of rhynchosaurs suggests that they were herbivores that fed on tough plants, possibly seed-ferns. *Hyperodapedon* has massive high claws on its feet which may have been used to dig up succulent tubers and roots by backwards scratching. The rhynchosaurs, when present, were dominant herbivores in their faunas, and yet they, like the dicynodonts died out 225 million years ago, at the Carnian-Norian boundary.

The third diapsid group, the prolacertiforms, appeared first in the mid Permian, and they radiated in the Triassic. Most of the Triassic forms probably looked like large lizards, but by the Mid Triassic, one of the most remarkable Triassic reptiles appeared, *Tanystropheus* from Central Europe. This reptile had a long neck, more than twice the length of the body. Young *Tanystropheus* have relatively short necks, and as they grew larger the neck

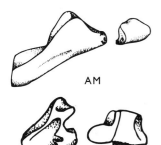

AM

CN

The archosaurs have two main ankle types (above). Crocodiles and their ancestor, collectively the crurotarsans, have a complex hinging ankle, the crocodile-normal (CN) type, while dinosaurs and their ancestors, the ornithodirans, have a simple hinge, the advanced mesotarsal (AM) type.

sprouted at a remarkable rate. Its function is a mystery. The sharp teeth suggest that *Tanystropheus* fed on meat, while the limbs and other features may indicate a life in the water. Indeed, many of the specimens are found in marine sediments, and *Tanystropheus* may have been a coastal swimmer that fed on small fishes which it caught by darting its head about rapidly.

The Ancestry of the Crocodiles

The archosaurs underwent a major phase of evolutionary diversification at the end of the Early Triassic. They branched into two major groups, one of which led to the crocodiles, and the other to the dinosaurs and birds. The crocodile line, the crurotarsans, all have a complex ankle joint, in which there is hinge-like movement, and some rotation between the two main ankle bones. The dinosaur line, the ornithodirans, have a simplified ankle joint which acts simply as a hinge, allowing the foot only to flap up and down, and not to rotate much.

The crurotarsans include a variety of important Late Triassic reptile groups, most of which died out at the end of the Triassic, but one group, the croco-

The phytosaur Parasuchus, *from the Late Triassic of India (above) was a fish-eater. It looks superficially like a crocodile, but the details of its skull show it evolved its long snout independently.*

diles, survives to the present day. One crurotarsan group, the phytosaurs, are so far known only from the Late Triassic of Germany, North America, and India. *Parasuchus* from India is 2.5 metres long and looks just like a crocodile. The long narrow jaws are lined with sharp conical teeth that interlock in such a way that *Parasuchus* could seize a rapidly darting fish and pierce it with the long teeth, and then hold it firm while it expelled water from the sides of its mouth before swallowing. The nostrils of *Parasuchus* are raised on a mound of bone just in front of the eyes (not at the tip of the snout as in crocodilians), so that it could have lain just below the surface of the water with only its nostril-mound showing. *Parasuchus*, like many modern crocodilians did not only hunt fishes in the water. Two specimens have been found with stomach contents of small tetrapods—the bony remains of prolacertiforms and a small rhynchosaur—which were probably seized on the river bank and dragged into the water.

The aetosaurs, a second crurotarsan clade, were the first herbivorous archosaurs, and they radiated nearly worldwide in the Late Triassic (see page 69). *Stagonolepis* from Scotland had a tiny head, a powerful heavy tail and short stout legs. The skull has a blunt upturned snout which may have been used as a small shovel to dig around in the soil for edible tubers and roots. The body is encased in an extensive armour of heavy bony plates that are set into the skin, a necessary defence against the major carnivores of that time, the rauisuchians.

Saurosuchus, a rauisuchian from the Late Triassic of Argentina, reached 6 or 7 metres in length. *Saurosuchus* had a specialised erect gait in which the femur (thigh bone) remained vertical and fitted from below into a nearly horizontal set of hip bones. This is very different from the erect posture of dinosaurs, birds, and indeed mammals, in which the femur has a ball-and-socket head that fits sideways into a vertical hip joint. The skeletons of *Saurosuchus* were found in association with a rich fauna of aetosaurs, rhynchosaurs, small and large mammal-like reptiles (dicynodonts and cynodonts), and some rare amphibians and small dinosaurs (see pages 70–71). These probably all formed part of the diet of *Saurosuchus*, but the rhynchosaur *Scaphonyx* was probably the main prey animal, since it was extremely abundant in the Ischigualasto fauna, and was large enough to make a succulent meal.

True crocodilians arose in the Early Jurassic, but there were a number of close relatives in the Late Triassic, all of which are included in the clade Crocodylomorpha. Some of the Late Triassic crocodylomorphs seem most uncrocodilian at first sight. An example is *Terrestrisuchus* from the Late Triassic of South Wales, a lightly-built, delicate 0.5-metre long animal. It has a long skull with slender pointed teeth, and long hind limbs that suggest it was a biped. It probably fed on small reptiles, insects, and other invertebrates. How can this fully terrestrial insectivorous biped be a close relative of the crocodilians? *Terrestrisuchus* has a number of diagnostic crocodilian characteristics. The main bones of the wrist are elongated into rod-shaped elements, instead of being button-shaped, as in other reptiles, and there are other crocodile-like features in the shoulder, hip girdles, and skull.

The phytosaurs, aetosaurs, and rauisuchians all died out right at the end of the Triassic. Only the basal stock of the crocodiles survived into the Jurassic.

In Triassic Seas

There were three main groups of reptiles in Triassic seas, the nothosaurs, placodonts, and ichthyosaurs. The nothosaurs were elongate animals with small heads, long necks and tails, and paddle-like limbs. They are best known from the Middle Triassic of Central Europe where animals like *Pachypleurosaurus* have been found abundantly in marine sediments. These were clearly mainly aquatic animals, which used wide sweeps of their deep tails to produce swimming thrust. The limbs may have been used to some extent in steering, but they were probably held along the sides of the body most of the time in order to reduce drag. The skull is long and lightly-built with pointed peg-like teeth. These suggest a diet of fishes which the agile nothosaurs could have chased and snapped up with darts of their long necks.

The placodonts, a very different group of marine reptiles, were also most abundant in the Middle Triassic of Central Europe, and disappeared during the Late Triassic. *Placodus* has unusual teeth: six spoon-like incisors at the front, and fourteen broad palatal teeth, which are flattened and covered with heavy enamel. They were clearly used in crushing some hard-shelled prey, probably molluscs, which were levered off the rocks with the incisors, smashed between the massive palatal teeth and the flesh extracted.

The ichthyosaurs (literally 'fish lizards') were the most obviously aquatic reptiles of all with their dolphin-like bodies—no neck, streamlined form, paddles, and fish-like tail. They arose in the Early Triassic and radiated in the Middle and Late Triassic. *Mixosaurus*, a typical Middle Triassic form from Central Europe, had advanced paddles with short limb bones and an excess number of finger and toe bones. Some specimens even preserve blackened traces of the body outline which

The early ichthyosaur Mixosaurus, *from the Middle Triassic of Europe, was a slender fish-eater.*

show that there was a tall dorsal fin in the middle of the back, a sickle-shaped tail fin, and mitten-like paddle coverings. Some Late Triassic ichthyosaurs reached lengths of 15m. They had long bullet-shaped heads, teeth only at the front of the snout, a vast rib cage, and tremendously elongated limbs.

The Origin of the Dinosaurs

The oldest dinosaurs, known especially from South America (see pages 70–71) date from the earliest Late Triassic (mid Carnian), some 228 million years ago. However, the closest relatives of dinosaurs, the lagosuchids, are latest Mid Triassic in age, and this implies an origin of dinosaurs by at least that time, perhaps 235 million years ago.

Marasuchus, the best-known lagosuchid, was a lightly-built flesh-eater, some 1.3 metres long, that presumably preyed on small fast-moving animals such as cynodonts and procolophonids, as well as perhaps worms, grubs, and insects. The skull is incompletely known, but the skeleton shows many dinosaur-like characteristics, such as a swan-like S-curved neck, an arm which is less than half the length of the leg, the beginnings of an opening in the hip joint, and other features of the hip and limb bones associated with fully erect posture. *Marasuchus* was clearly a biped, running on its hindlimbs, and the long tail was presumably used as a balancing organ. It may have used its hands for grappling with prey and for passing food to its mouth.

Marasuchus, *a slender lightweight predator from the Middle Triassic of Argentina, was a close relative of the dinosaurs.*

There has been a long-running debate about the oldest dinosaur remains. Part of the problem was that no-one could decide just what a dinosaur was, and for most of this century, up to the 1980s, most palaeontologists believed that the dinosaurs were a chance assemblage of large reptiles that had arisen from three or more separate ancestors (see pages 72–73). Older accounts frequently state that the dinosaurs arose early in the Triassic, and they often quoted evidence in the form of skeletons and footprints. However, the supposed skeletal remains of dinosaurs from before the Late Triassic turn out to belong to prolacertiforms, rauisuchians, and other non-dinosaurian groups. Dinosaur footprints, which have a characteristic three-toed appearance (because toes one and five are either short, or absent) had also been recorded from the Early and Mid Triassic of various parts of the world, but critical re-examination has now shown that they have been wrongly identified.

The oldest true dinosaurs are known from the early part of the Late Triassic (the Carnian Stage, 230–225 Myr) from various parts of the world. The best specimens come from the Ischigualasto Formation of Argentina, source also of the rauisuchid *Saurosuchus.* The Ischigualasto dinosaurs, *Eoraptor* and *Herrerasaurus*, are relatively well known from nearly complete specimens (see pages 70–71), and they give an insight into the days before the dinosaurs rose to prominence: *Eoraptor* and *Herrerasaurus* were minor components of their faunas.

Dinosaurs are known from the Carnian of North America, where forms such as *Coelophysis* (see pages 74–75) appeared towards the end of that stage, and then appeared to become much more abundant in the Norian stage (225–205 Myr). In the Norian, other dinosaur groups radiated, and some of them, such as the herbivorous prosauropod *Plateosaurus* from Central Europe (see pages 76–77) became abundant and large. The largest prosauropods reached lengths of 5–10 metres in the Late Triassic.

Competition or Mass Extinction

D*id the dinosuars succeed by driving other reptiles to extinction, or were they merely lucky?*

"Dinosaurs appeared as the dominant land vertebrates only after this great disappearance of great numbers of therapsids and of many archosaurs and archosauromorphs. Thus, the initial radiation of dinosaurs was done in an ecological near-vacuum..."
David E. Fastovsky and David B. Weishampel, 1996

The seed from Glossopteris *existed in Gonwanaland in the Late Permian, and was related to* Dicroidium, *a dominant Triassic plant.*

There are currently two ways of viewing the radiation of the dinosaurs in the Late Triassic. Either they 'took their chance' after a mass extinction event and radiated opportunistically, or they competed over a longer time-span with the mammal-like reptiles and eventually prevailed.

Until recently, most palaeontologists assumed that the competitive model was correct for four reasons. Firstly, many considered that the dinosaurs had arisen from many different ancestors. Secondly, the origin of the dinosaurs was seen as a drawn-out affair, that started well-down in the Middle Triassic, and involved extensive and long-term competition. The dinosaur ancestors were regarded as superior animals, with advanced locomotory adaptations (erect gait) or physiological advances (warm-bloodedness, or cold-bloodedness: both cases have been argued!) which progressively competed with, and caused the extinction of, all of the mammal-like reptiles and basal archosaurs. Thirdly, the appearance of the dinosaurs has often been regarded as a great leap forward in evolutionary terms.

A fourth reason why many palaeontologists accepted the competitive model for the radiation of the dinosaurs was more general. It has commonly been assumed that the evolution of life is in some way progressive, that more recent plants and animals are inevitably better than those that went before. So, modern mammals might be said to be better competitors than archaic mammals, archaic mammals might be better than dinosaurs, and dinosaurs might be better than their forerunners. This assumption of progress has never been demonstrated, and indeed the major changes in world floras and faunas might equally well be associated with expansions into new ecospace, with no direct competition with pre-existing forms at all.

When I studied this question, as part of my PhD research, I found several lines of evidence that the dinosaurs had radiated after an extinction event:

(1) The pattern in the fossil record does not support the competitive model. Dinosaurs appeared in the mid Carnian (228 Myr ago), or earlier, and they were rare elements in their faunas (less than 5% of individuals) until a major extinction event at the end of the Carnian (225 Myr ago), when various fam-

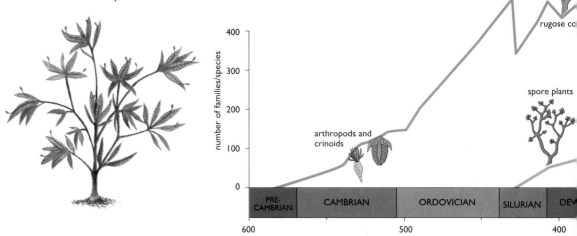

ilies of basal archosaurs, mammal-like reptiles, and the rhynchosaurs died out. The key extinctions were those of the dominant herbivore groups, the dicynodonts, herbivorous cynodonts, and rhynchosaurs. Dinosaurs radiated during the Norian stage (225–205 Myr ago), and all the major lineages (see pages 72–73) appeared during this time. In Norian faunas, dinosaurs represent 50–90% of individuals, a dramatic increase from their pre-Norian totals. Dinosaurs diversified further in the Early Jurassic after a second mass extinction at the very end of the Triassic when the remaining basal archosaurs, and other groups, died out.

(2) The 'superior adaptations' of dinosaurs were probably not so profound as was once thought. Many other archosaurs also evolved erect gait in the Late Triassic, and yet they died out (e.g. aetosaurs and some early crocodylomorphs). The physiological characteristics of dinosaurs—whether they were warm-blooded or not, for example—cannot be determined with confidence.

(3) There is good evidence for other extinctions at the end of the Carnian. The *Dicroidium* flora of the southern hemisphere gave way to a worldwide conifer flora about this time. There were also turnovers in marine communities, particularly in reefs, and there was a shift from pluvial (heavy rainfall) climates to arid climates throughout much of the world. These changes may have caused the extinctions of the dominant herbivorous tetrapods.

(4) The idea that simple competition can have major long-term effects in evolution is probably an over-simplification of a complex set of processes. That families or orders of animals can compete with each other is very different from the ecological observation of competition within or between species. In palaeontological examples like this, competition has often been assumed to have been the mechanism, the evidence has generally been weak.

New work on dinosaur origins has given a very full picture of the nature of the oldest dinosaurs. Clade Dinosauria evidently originated from a single ancestor 230–235 million years ago, probably a small bipedal hunter of some kind, perhaps in South America. Dinosaurs were some of the most successful animals that ever lived, but like the mammals which came after, perhaps they owe their success more to luck than to any major competitive process.

The diversification of life (below) can be traced from the fossil record of marine invertebrates, land plants, and land tetrapods. The end-Permian mass extinction was the biggest of all time; but a smaller event 225 Myr ago, in the Late Triassic, triggered the rise of the dinosaurs.

Diversity and time

— marine invertebrate families

— vascular land plant species

— land tetrapod families

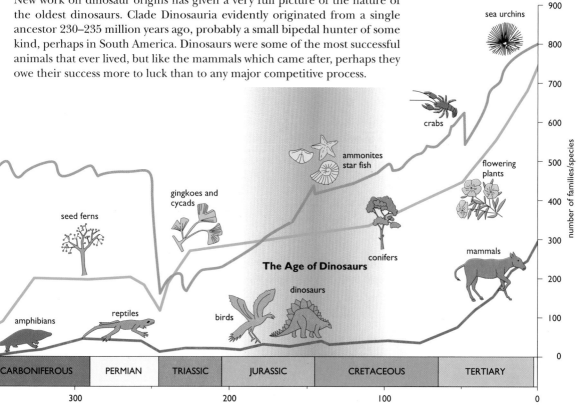

Timeline II

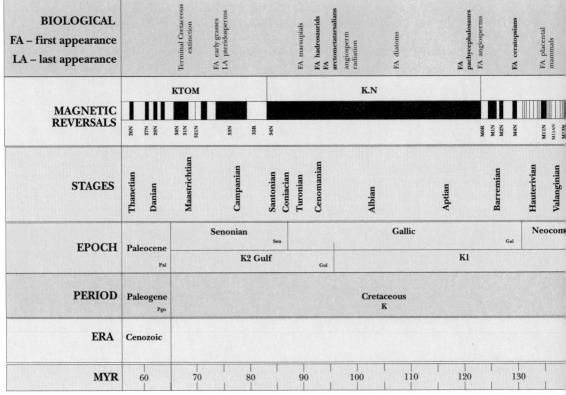

BIOLOGICAL FA – first appearance LA – last appearance		

In the time scale (above), some key biological events are noted. The divisions of time here are based on three kinds of evidence (see pages 18-19). First, the scale of magnetic reversals shows times when the Earth's magnetic field changed from normal (N), as it is today (shown black) to reversed (R, shown white), when the magnetic north and south poles flipped over. The magnetisation is measured from grains of iron locked into ancient rocks. The stages, epochs, periods, and eras are based on relative dating using fossils, and these indicate the international time scale. The exact ages, in millions of years (MYR) are based on measurements of radioactive materials in the rocks and on calculations of decay rates.

CLASSIFICATION OF THE DINOSAURS
Superorder Dinosauria
 Order Saurischia
  Suborder Theropoda
   Family Herrerasauridae
   Infraorder Ceratosauria
   Infraorder Tetanurae
    Family Megalosauridae
    Family Allosauridae
    Division Maniraptora
     Family Dromaeosauridae
     Class Aves (birds)
     Family Oviraptoridae
     Subdivision Arctrometatarsalia
      Family Troodontidae
      Family Ornithomimidae
      Family Tyrannosauridae
  Suborder Sauropodomorpha
   Infraorder Prosauropoda
    Family Thecodontosauridae
    Family Plateosauridae
    Family Melanorosauridae
   Infraorder Sauropoda
    Family Vulcanodontidae
    Family Euhelopodidae
    Division Neosauropoda
     Family Cetiosauridae

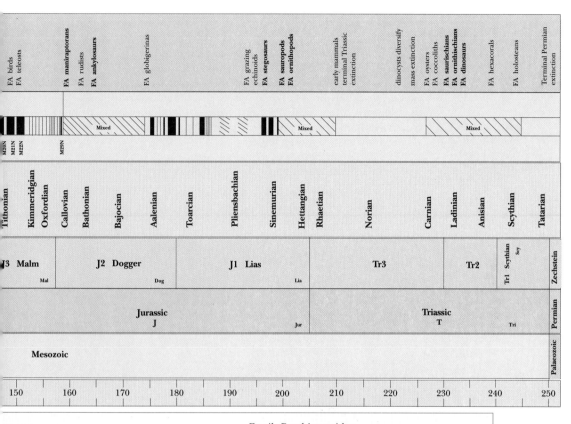

Family Brachiosauridae
Family Camarasauridae
Family Titanosauridae
Subdivision Diplodocoidea
     Family Nemegtosauridae
     Family Diplodocidae
Order Ornithischia
     Family Pisanosauridae
     Family Fabrosauridae
     Suborder Thyreophora
          Family Scelidosauridae
          Infraorder Stegosauria
          Infraorder Ankylosauria
               Family Nodosauridae
               Family Ankylosauridae
     Suborder Cerapoda
          Infraorder Pachycephalosauria
          Infraorder Ceratopsia
               Family Psittacosauridae
               Family Protoceratopsidae
               Family Ceratopsidae
          Infraorder Ornithopoda
               Family Heterodontosauridae
               Family Hypsilophodontidae
               Family Iguanodontidae
               Family Hadrosauridae

The Triassic World

The Triassic period in which dinosaurs originated was a time of aridity, when plants and animals were very similar worldwide.

Kuehneosaurus, *an extraordinary gliding animal of the Late Triassic period, used a membrane stretched over its elongated ribs as a kind of parachute. When resting in a tree,* Kuehneosaurus *folded its ribs neatly away along the side of its body.*

The Triassic was a time of arid-climate plants and abundant reptiles. The continents reached their maximum phase of fusion about 230 million years ago, during the Late Triassic. Continental plates had manoeuvred themselves into a position where the southern continents (Gondwanaland) were in extensive contact with the northern continents (Laurasia), forming the supercontinent Pangea. Parts of China and central Asia may still have formed separate islands, but most parts of the Earth were in contact. When this phase ended, rifting began between Africa and North America, and the continents began to move apart towards their present positions.

Much of the Triassic world lay in the equatorial belt. Climates were hot and monsoonal in much of North America, Europe, Africa, India and Australia. There were no polar ice caps, and the temperature gradient from the Equator to the poles was much less than it is now. In the last half of the Triassic, climates became drier, perhaps because of changes to circulation patterns brought about by the final closure of the oceans to form Pangea.

The hot climate favoured certain kinds of plants and animals. The main vegetation types were seed ferns and conifers, which were adapted to low or variable water conditions, and massive horsetails in damper situations. Land animals included all kinds of insects, spiders, earthworms, snails and reptiles. The reptiles fed on the tough plants and captured insects and worms in the undergrowth. Some even took to the air as gliders or as powered flyers. True mammals appeared late in the Triassic. In rivers and lakes lived many kinds of shellfish, as well as thick-scaled fish. These were preyed upon by broad-headed amphibians and phytosaurs, which are distant relatives of the crocodile. In the seas, the dolphin-like ichthyosaurs and long-necked nothosaurs fed on fish.

The Triassic world, 250–205 million years before present

	ancient continent
	ancient continental shelf
	ancient mountain chain
	cold ocean current
	warm ocean current
	continent movement
	modern coastline

NORTH AMERICA

The first mammals such as Megazostrodon appeared in the Late Triassic. They were tiny shrew-like creatures that fed on insects and probably only ventured out at night.

Marine reptiles of the Triassic included nothosaurs (below), with their long necks and sharp teeth for spearing fish, and the dolphin-like ichthyosaurs (far right), which also hunted fish. Ichthyosaurs may have also preyed on shellfish.

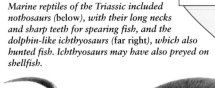

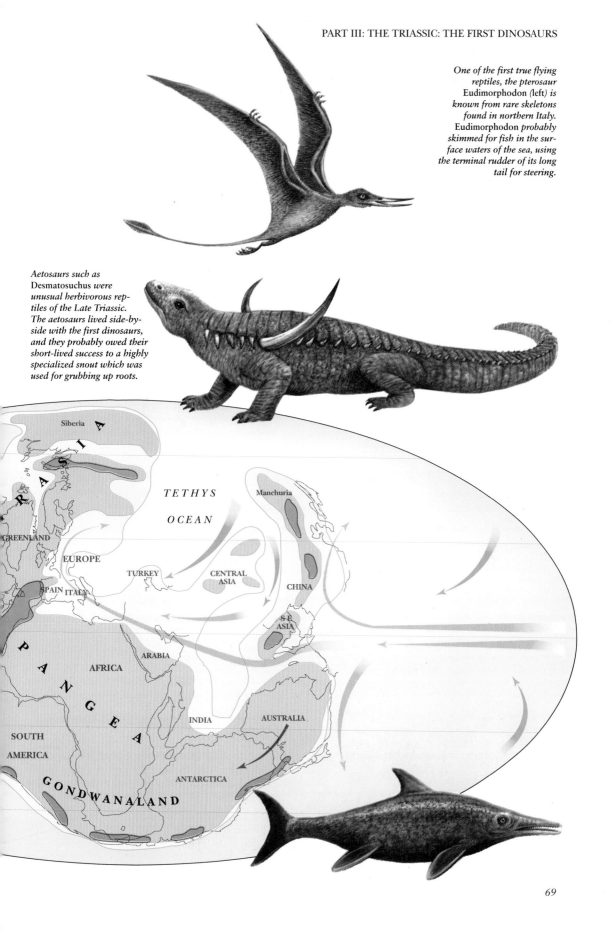

One of the first true flying reptiles, the pterosaur Eudimorphodon *(left) is known from rare skeletons found in northern Italy.* Eudimorphodon *probably skimmed for fish in the surface waters of the sea, using the terminal rudder of its long tail for steering.*

Aetosaurs such as Desmatosuchus *were unusual herbivorous reptiles of the Late Triassic. The aetosaurs lived side-by-side with the first dinosaurs, and they probably owed their short-lived success to a highly specialized snout which was used for grubbing up roots.*

The First Dinosaurs

The first dinosaurs were small bipedal carnivores from Argentina. These early forms soon spread worldwide.

The skull of Herrerasaurus *shows its meat-eating adaptations: powerful jaws lined with long sharp recurved teeth. These are ideal for seizing and holding struggling prey. The prey can move only one way because of the recurved teeth: down into the throat. The specialised joint in the middle of the lower jaw suggests that* Herrerasaurus *may be a member of the Theropoda, the great group of carnivorous dinosaurs.*

New studies of specimens from Argentina have confirmed that the dinosaurs arose at the beginning of the Late Triassic, during the Carnian Stage, perhaps 230 million years ago. The bones of slender bipedal dinosaurs were found in the Ischigualasto Formation of Argentina in the 1950s and 1960s, and named *Herrerasaurus*. Remains of *Herrerasaurus*, and of some other early dinosaurs, were sparse until the 1990s when a new wave of collecting in the Carnian-age Ischigualasto Formation began. Argentinian palaeontologist Fernando Novas and North American Paul Sereno mounted several expeditions to the Ischigualasto Valley, the 'valley of the moon', and they unearthed a complete skeleton of *Herrerasaurus*.

The new materials of *Herrerasaurus* confirmed that this was a dinosaur, measuring 3–6 metres in length as an adult, and with fully dinosaurian hindlimbs. Further dinosaurian characteristics included features of the backbone and of the forelimb. In the pelvic area, most reptiles have two specialised vertebrae, the sacral vertebrae, that support the hip bones: dinosaurs have three or more, and indeed the new specimens of *Herrerasaurus* show that it had three sacral vertebrae. In addition, the humerus (upper arm bone) of *Herrerasaurus* has an elongate crest running down the front margin, again a feature seen only in dinosaurs. There is a dispute, however, about how much of a dinosaur *Herrerasaurus* truly is: is it more primitive than all subsequent dinosaurs, or can it be assigned to the Theropoda, the group of carnivorous forms? Sereno has championed the idea that *Herrerasaurus* is a theropod based on its possession of a specialised joint in the lower jaw, seen only in those meat-eating dinosaurs.

Sereno and Novas made a further spectacular discovery during their collecting in the 1990s, the skeleton of another, perhaps more primitive, dinosaur, which they named *Eoraptor*, or 'dawn hunter'. *Eoraptor* was much smaller than *Herrerasaurus*, being about 1 metre long. It had all the dinosaur characteristics seen in *Herrerasaurus*, including a strong three-fingered hand which was presumably used for grappling with prey. However, the skull of *Eoraptor* represents a basic dinosaur design, with fewer specialisations hinting at Theropoda, or any of the other major dinosaur groups.

Herrerasaurus and *Eoraptor* were agile hunters that could run fast, and manoeuvre effectively. They had the advantage of speed and of having arms that were free from locomotory functions. They could prey on the other animals of Ischigualasto times, the smaller cyn-

Recnstruction of the skeleton of Herrerasaurus, *based on the new skeleton discovered by Novas and Sereno. This 3-6 metre long dinosaur has powerful hindlimbs, a long tail for balance, and strong arms for wrestling with prey. It probably ran with its backbone nearly horizontal, in which position it is perfectly balanced over its hindlimbs.*

odonts and lizard-sized animals, as well as juvenile rhynchosaurs. Top carnivores were still the rauisuchians, a group of basal archosaurs. New studies of the ecology of Ischigualasto times show that these early dinosaurs were a minor part of their communities, representing perhaps 5% of all animals. The age of the dinosaurs had begun, but in rather a discrete manner!

Eoraptor, *perhaps the most
primitive dinosaur* (right),
*from the Carnian-age
Ischigualasto Formation of
Argentina, was an active
hunter which fed on small
lizard-sized reptiles. Eoraptor
was the quickest hunter of its
day, and was able to exploit
the small animals, even the
warm-blooded cynodonts,
closely related to the later
mammals. There are
extensive sequences of Triassic
rocks in Argentina* (below),
*running in a strip down the
eastern flanks of the Andes
Mountains. The Ischigualasto
Basin lies in the north of the
country, and fossil reptiles
have been found in several
other localities.*

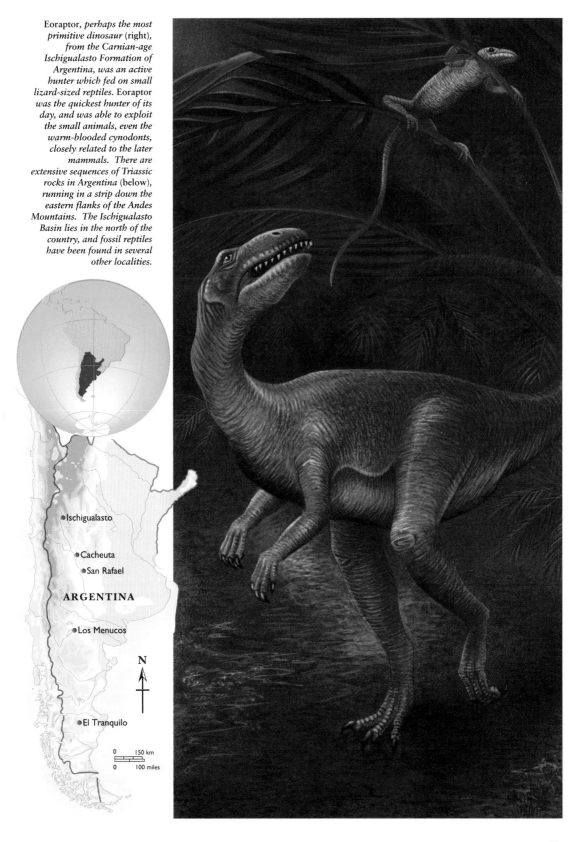

Ischigualasto

Cacheuta

San Rafael

ARGENTINA

Los Menucos

El Tranquilo

N

| 0 | 150 km |
| 0 | 100 miles |

Dinosaur Phylogeny

Cladistic analysis has solved long-term disputes about the evolution of the dinosaurs: the group is monophyletic, and it evolved from bipedal meat-eating forms.

Ever since the discovery of the dinosaurs, palaeontologists have debated their true relationships. They were clearly reptiles, and Victorian anatomists identified their closest living relatives as crocodiles and birds. This posed a conundrum, since crocodiles and birds seem so utterly different from each other. However, both of these living groups actually share many features of their skulls and skeletons, and molecular evidence confirms that they belong to a wider group, the Archosauria. But where do the dinosaurs fit?

Sir Richard Owen had the right idea in 1842 when he invented the name Dinosauria: he regarded this as a monophyletic group, in other words, a group which had originated from a single ancestor, and which included all the descendants of that ancestor. T. H. Huxley (1825–1895) solved the issue of the relationships of dinosaurs and birds about 1870, when he noted the extraordinary resemblances between some of the small bipedal meat-eating dinosaurs and the recently discovered oldest bird, *Archaeopteryx*.

In 1887, Harry Seeley noted that dinosaurs fell into two great groups, the Saurischia ('reptile hips') and Ornithischia ('bird hips'), based on their pelvic structures. Saurischians show the classic reptilian hip structure, with the two lower hip bones, the pubis and ischium pointing respectively forwards and backwards. Ornithischians, on the other hand, have a modified hip structure, in which the pubis has swung backwards to run parallel to the ischium. Most palaeontologists then assumed that the Saurischia and Ornithischia were entirely independent groups, and that the word 'dinosaur' could be used only in a loose sense to refer to a motley array of large Mesozoic reptiles.

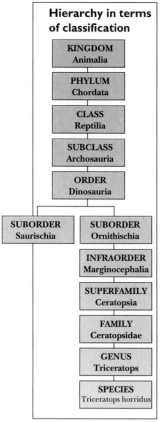

Hierarchy in terms of classification

- **KINGDOM** Animalia
- **PHYLUM** Chordata
- **CLASS** Reptilia
- **SUBCLASS** Archosauria
- **ORDER** Dinosauria
- **SUBORDER** Saurischia
- **SUBORDER** Ornithischia
- **INFRAORDER** Marginocephalia
- **SUPERFAMILY** Ceratopsia
- **FAMILY** Ceratopsidae
- **GENUS** Triceratops
- **SPECIES** Triceratops horridus

Classification of a dinosaur (above), the well-known horned plant-eater Triceratops horridus *(see pages 132-133). The species* horridus *is one of several in the genus* Triceratops. Triceratops *and other genera make up the family Ceratopsidae in the superfamily Ceratopsia, and they, together with the pachycephalosaurs and others, make up the infraorder Marginocephalia. The sequence of inclusive categories continues up to the Kingdom Animalia. This inclusive classification scheme, reflects the pattern of evolution (right).*

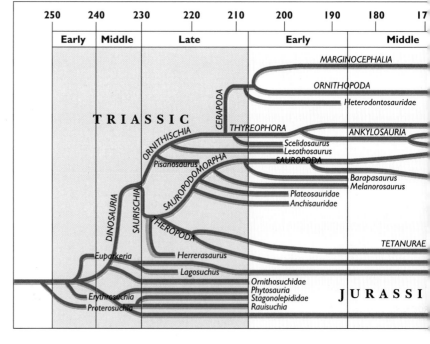

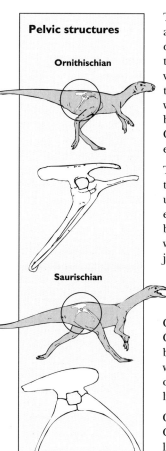

Pelvic structures

Ornithischian

Saurischian

The problems multiplied early in the 20th century, when further divisions among the Saurischia and Ornithischia were established. The Saurischia is divided into Theropoda, the meat-eating dinosaurs, and Sauropodomorpha, the large long-necked plant-eaters. The Ornithischia is also subdivided into various subgroups, the two-legged Ornithopoda, the horn-faced Ceratopsia, the plate-spined Stegosauria, and the armoured Ankylosauria. Perhaps there were more than two fundamental groups of dinosaurs, and perhaps there had been three or more origination events somewhere low in the Triassic? Confusion peaked in the 1960s and 1970s, when some palaeontologists hinted at four or five independent origins.

The problem not faced by these palaeontologists was the long list of characteristics shared by all dinosaurs. The key features are associated with the upright bipedal posture of the oldest dinosaurs (see pages 70–71). Unlike earlier archosaurs, dinosaurs stood erect, with their limbs tucked under their body, instead of in a sprawling posture, as in their ancestors. Sprawlers walk with the limbs angled out at the side of the body, and the hip, knee, and ankle joints are all complex rotatory hinges. Erect animals have straight up and down legs. The femur (thigh bone) fits into the hip bones from below with a ball and socket joint, and the knees and ankles are simple hinges with no significant sideways rotation.

Cladistic analysis (see pages 26–27) in the 1980s resolved the problem. Cladistics allows for no vagueness: firm decisions about relationships have to be made, and it turned out that the evidence for a monophyletic Dinosauria was enormous when compared with any other hypothesis that the dinosaurs consisted of many separate lines, each of which had originated independently somewhere among the basal archosaurs.

Cladistic analyses have confirmed the existence of the Saurischia and the Ornithischia. The Saurischia are defined, not by their possession of a 'lizard-like' pelvis, since that feature is primitive, shared with all other reptiles, but by other featutres of the skull, backbone, and hand.

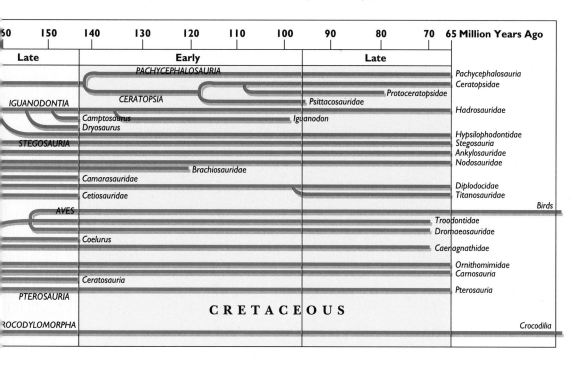

Coelophysis from Ghost Ranch

A mass accumulation of skeletons of the small dinosaur Coelophysis, found in 1947, was the first evidence for herding behaviour.

Ghost Ranch

The first Triassic dinosaurs were found in Europe in the 1830s, but little was known from North America until the 1880s. Little, that is, other than footprints. The Triassic sediments of the Connecticut Valley in the eastern United States were a rich source of three-toed footprints, some small and some large. The first specimen was found in Connecticut in 1802. More specimens were collected in the 1830s. We recognise now that these Connecticut Valley footprints were made by a variety of meat-eating and plant-eating dinosaurs.

In 1881, Edward Drinker Cope (see pages 24–25) described a new small dinosaur, based on some pieces of the backbone, some rib fragments, hip bones, and limb bones, which had been found in the Late Triassic sediments of north-western New Mexico. Cope recognised the animal as a carnivorous dinosaur, and he named it *Coelophysis* ('hollow form'), since many of the bones were hollow inside, just as in birds.

Coelophysis skeletons have been found in various parts of North America (above), particularly in the south-west, in Arizona and New Mexico, but also in the Connecticut valley in the east, where footprints are also abundant. The most spectacular collection of over 100 skeletons was made in 1947 close to the Arroyo del Yeso in north-west New Mexico, on the lands of Ghost Ranch (below).

Then in 1947, an expedition from the American Museum of Natural History, which was working in the fossiliferous Late Triassic redbeds of New Mexico, came upon an astonishing bone bed. While working in the Arroyo Seco ('dry canyon'), very close to the location of Cope's bone scraps, they found a mass accumulation of complete skeletons of *Coelophysis* . The locality lay on the lands of Ghost Ranch, and that is the name used to refer to this famous locality. The American Museum team gradually worked back into the wall of the canyon, uncovering more than a hundred skeletons of *Coelophysis*, some small, some large.

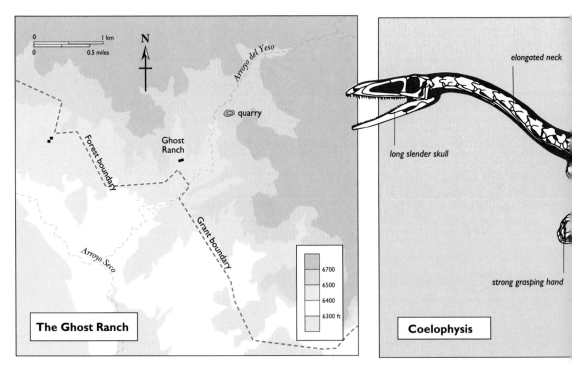

The Ghost Ranch

Coelophysis

elongated neck

long slender skull

strong grasping hand

The excavators had a huge logistical problem in securing and protecting so many delicate skeletons, and in transporting them back to New York. Thirteen major blocks were shipped out. Later joint expeditions in 1981–82 and 1985 extended the site, and a further 16 blocks were removed. One of the palaeontologists who was present on the 1947 expedition, Ned Colbert, was able to produce a full description of *Coelophysis*, based on the wealth of specimens, in 1989.

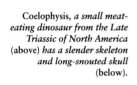

How could so many skeletons of one species be concentrated in such a restricted area (the whole quarry area is about 6 x 20 metres in extent)? Perhaps a whole herd was overwhelmed by some catastrophe, or perhaps the animals were concentrated by sedimentary processes after their death. The sediments in which the fossils were found are sandstones and mudstones deposited by ancient river systems. Ned Colbert prefers the mass-death model, in which a herd, including young and old animals, were perhaps crossing a river, and were overwhelmed by a flood. The skeletons are mainly complete and well preserved. Some are disarticulated, meaning that parts of the body have drifted away, so the bodies were scavenged and decomposed, and perhaps disturbed by water currents, before they were finally buried. These ideas require further checking of the sediments.

Coelophysis, a small meat-eating dinosaur from the Late Triassic of North America (above) has a slender skeleton and long-snouted skull (below).

The American Museum collection shows considerable variation in size. Skull lengths range from 8 to 26 centimetres, so clearly a range of juveniles and adults are present. These skull lengths scale up to body lengths of 0.8–3.1 metres. No babies have yet been identified, however, since they would have had even smaller skulls, perhaps only 2–3 centimetres long, and body lengths of less than 20 centimetres. Colbert was also able to identify possible sexual dimorphism: males seem to have larger skulls, longer necks, and shorter arms than females. Of course, the sex cannot be identified with certainty, but at least there appear to be two forms among the adult *Coelophysis*.

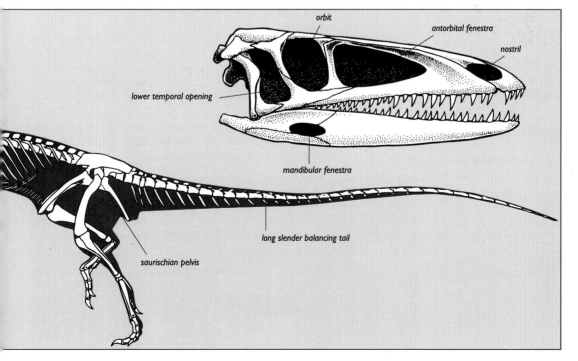

orbit

antorbital fenestra

nostril

lower temporal opening

mandibular fenestra

long slender balancing tail

saurischian pelvis

Plateosaurus from Trossingen

Friedrich von Huene (1875–1969), one of the most prolific vertebrate palaeontologists of this century (above). He began his studies in the late 19th century, working on fossil brachiopods, but by 1902 had shifted his interests to fossil reptiles. He carried out major collecting trips in his native Germany, but also in previously unexplored parts of South America and Africa. He also toured the museums of England, France, and North America, and in a long research career of nearly 70 years, published hundreds of descriptions of fossil reptiles.

Recent excavation at a Plateosaurus locality (below right), showing the use of a bulldozer to shift a massive block containing bones. Trossingen is one of forty or more Plateosaurus localities in Germany and neighbouring parts of Switzerland (below).

Plateosaurus *was the first really large dinosaur. Huge numbers of skeletons have been found in sandstone quarries in southern Germany and Switzerland.*

The first Triassic dinosaur, the small herbivore *Thecodontosaurus*, was reported in 1836 from England, and a year later followed the announcement of *Plateosaurus*, an even larger form, from the Late Triassic Keuper beds of south-west Germany. During the remainder of the 19th century, dozens of new specimens of *Plateosaurus* came to light in the sandstone quarries around Stuttgart and Tübingen, in the Lande of Baden-Württemberg, and just across the border, in Switzerland. The particular parts of the German Keuper that yield *Plateosaurus* skeletons, the Stubensandstein and the Knollenmergel, are dated as Norian, an interval of time that lasted from about 225 to 205 million years ago.

In 1907–8, Baron Friedrich von Huene published a huge monograph describing *Plateosaurus* and its allies. Von Huene opened an excavation at Trossingen, 50km south of Tübingen in the sides of a small wooded valley, in the summer of 1921. He employed an army of museum technicians and enthusiastic students, and they began to clear the rock from above the bone layer. Huge skeletons and partial skeletons were uncovered, and after a summer of hard work, many thousands of bones were encased in plaster, and taken north to the university museum at Tübingen. The work continued in 1922, and further excavations in the early 1930s, produced yet more bones. In all, the site yielded nine complete skeletons, 17 incomplete skeletons, and isolated remains of another 40 individuals.

Von Huene was clear about the origin of the deposit. He believed that the Trossingen site represented the site of death of a herd of *Plateosaurus* who were traversing a wide desert, some 100 kilometres across, walking from one water body to another during the dry season. More recent sedimentological study at Trossingen has shown that the Keuper sediments are clearly water-laid, and indicate deposition in lakes and rivers. However, there are two explanations for the skeleton accumulation. David Weishampel of Johns Hopkins University has argued that the accumulation of skeletons is secondary, and that the animals died elsewhere, and were washed together on to a sand bank where the carcasses piled up. Martin Sander of the University of Bonn argues, on the other hand, that the animals died in situ, that they became mired in mud pools and were trapped. He notes that many of the skeletons were preserved with their bellies downwards, which he interprets as evidence that they died in a standing position.

Plateosaurus (right),
has peg-like teeth
and the hands had
huge thumb claws,
used perhaps to gather in
plant material from tall trees.
Herds of Plateosaurus (below)
fed in the lush woodlands of
central Europe in the Late
Triassic, and their close relatives ranged over southern
Africa and China.
Plateosaurus is a prosauropod,
a basal sauropodomorph.

IV: The Jurassic: heyday of giant dinosaur

The Jurassic period witnessed the origin of the armoured dinosaurs, the stegosaurs and ankylosaurs, and the origin and radiation of the giant sauropod plant-eaters. This was a time of lush forests and humid climates, and dinosaurs wandered on to all continents.

Ammonites in limestone from the Lower Triassic sediments of Robin Hood's Bay, Yorkshire, on the east coast of England. Ammonites were abundant and diverse in the Jurassic, and they are critical in dating rocks of the Jurassic.

The Jurassic period (205–145 Myr) was an important time in the evolution of the dinosaurs, and indeed in the evolution of modern life. The dinosaurs had already existed for 25 million years in the Late Triassic (see pages 63–65, 70–77), and the plant-eating prosauropods and active flesh-eating ceratosaurs were already well established. Other dinosaur groups, the ornithischians, and perhaps the tetanuran meat-eaters, had also appeared, but were much rarer.

There was a major mass extinction event at the end of the Triassic period, when many marine groups were wiped out. The ammonoids, coiled shellfish, virtually disappeared. The conodonts (see page 42), common throughout the Palaeozoic, finally disappeared, as did many lines of brachiopods (see page 40), and other other seabed-dwellers. Among the reptiles, the last of the basal archosaurs, the phytosaurs, aetosaurs, and rauisuchians all died out at the same time, and this has led some reptile palaeontologists recently to claim that it was the terminal-Triassic mass extinction that triggered the radiation of the dinosaurs. These palaeontologists point to the importance of the marine extinctions, and they support the idea that the mass killing was caused by the impact of an asteroid, as for the KT event (see pages 134–135).

The idea is not well-supported. There is only limited evidence for impact: indeed the candidate crater, at Manicouagan in British Columbia, is more than 10 Myr too early in date. In addition, there is no evidence that dinosaurs suddenly diversified at the beginning of the Jurassic. Certainly, the last of the basal archosaurs disappeared, and the extinction of the rauisuchians in particular freed up the large carnivore niches. But the key extinctions had happened 20 Myr earlier, when the formerly dominant herbivores, the dicynodonts and rhynchosaurs, had died out, and the prosauropods and ceratosaurs radiated (see pages 64–65).

The Jurassic was an important time for life in the sea, as well as on land, and some new groups came on the scene that were important in the evolution of the giant reptiles, and in the origins of modern groups.

Ammonites and Life on the Seabed

The commonest and most evocative Jurassic marine fossils are ammonites. These coiled shells are found often in huge numbers in Jurassic sections, and they are often exquisitely preserved in their original calcite, or replaced by pyrite or other minerals. The design of the shell, an expanding coil from the centre, and the detail of their internal suture lines and external ornament, are aesthetically appealing to many people. Rock hounds pick them up in huge numbers from classic Jurassic marine rock outcrops, and they are popular items in rock shops, where people buy them as display items.

The true ammonites are characteristic of the Jurassic and Cretaceous periods, and they died out at the end of the Cretaceous. They are a group of the ammonoids, cephalopod molluscs, whose modern relatives include the *Nautilus*, which shares the expanding coiled shell shape, and squid and octopus. Ammonoids arose during the early Palaeozoic, and they existed formerly in coiled and straight forms during the Palaeozoic, but the group virtually disappeared during the end-Triassic mass extinction. Only two or three lines survived into the Jurassic, but these radiated rapidly and diversified into many forms. Ammonites measure typically a few centimetres across, but some reached giant sizes, up to 2m across in some latest Jurassic forms (the appropriately named *Titanites titanites*), and others were tiny, less than 1cm in diameter.

Cross section of an ammonite shell, showing the inner chambered structure. As the animal grew in size, it built ever-larger living chambers and closed off the old chambers with calcite barriers. These spaces have been filled with minerals during fossilization.

The ammonite shell was made up from a series of hollow chambers that were laid down as the animal grew larger. The animal itself was a fleshy octopus-like creature, equipped with large eyes, a horny beak for feeding, and many tentacles. The body occupied the last quarter- or fifth-whorl of the shell, and the tentacles could be withdrawn, and covered with a horny lid if danger threatened. Normally, the ammonite was in perfect equilibrium with the water and it floated suspended at any level. It could move rapidly by squirts of water from its siphon, which sent it shooting backwards. Ammonites fed on small organisms in the water, and they themselves were preyed on by shell-crushing fishes and reptiles. Indeed, some ammonite shells have been found with lines of puncture marks produced by the teeth of large reptiles.

The embryo ammonite shell is preserved at the centre of the coil, and the entire growth record of an individual animal can be seen in its shell. As the animal increased in size, it required more space, and the diameter of the coil increases constantly. From time to time, the creature had to quit its former quarters, move forward into a newly-created part of the shell tube, and build a calcite wall behind it. In this way, a series of closed chambers was left behind in the centre of the shell, and these were filled with air and water. A fleshy strand of tissue was maintained through the old abandoned chambers, and the ammonite could adjust the air-water mix to make itself neutrally buoyant at whatever level it found itself in the water column.

Ammonites have had an additional value to the geologist, apart from their biological importance as a major part of Jurassic and Cretaceous marine communities. They are key stratigraphic tools. From the early days, when geologists sought to make sense of the great thicknesses of sediment they saw around them, ammonites have been used for dating Jurassic and Cretaceous rocks (see pages 18–19). Ammonite species are easy to identify by shell size, external patterns, and suture lines. The suture lines, in particular, are very helpful. These intricate patterns of curves and branching bush-like shapes reflect the shape of the dividing walls between chambers of the shell, and in the true ammonites the suture lines are intricate, and they vary from species to species. Individual ammonite species are often widespread, being found in many oceans at the same time, and they evolved fast. This allows stratigraphers to divide up Jurassic time into remarkably short segments, in some cases as little as 200,000 years, which gives extraordinary information on rates of deposition and of evolution.

Ammonites lived side by side with a wide range of other animals in Jurassic seas. Close relatives were the belemnites, which are represented today by bullet-like fossil shells made from calcite, generally 10–30 cm long. These were internal shells, something like the cuttle 'bone' from a cuttlefish, that were located inside a squid-like animal. In parts of the Jurassic and Cretaceous, belemnites are abundant, and also used for dating.

On the seabed were increasing numbers of bivalves, some of them on sandy surfaces, others such as the oysters, which were new arrivals, attached to the rocks, and others burrowing into the sediment for protection. Gastropods, the spire-shelled molluscs, preyed on the bivalves and other shelled prey. New modern groups of corals, which had arisen in the Late Triassic, formed reefs, and these provided a home for diverse groups of animals. Some of them, like the crinoids or sea-lilies, fixed themselves to hard parts of the seabed, and filterfed. Others such as the gastropods and early shrimps and crabs moved about in search of food. Fishes and marine reptiles too were seen in increasing abundance and diversity.

Sharks and Bony Fishes

The first sharks are known from the Late Devonian, and the cartilaginous fishes radiated initially during the Carboniferous (see page 43). A second shark radiation, of the hybodonts, took place in the Triassic and Jurassic. *Hybodus* was a fast-swimming fish, probably very agile and capable of accurate steering using its large pectoral (front) fins. The hybodonts had a range of tooth types, from triangular pointed flesh-tearing teeth to broad button-shaped crushers, for dealing with hard-shelled crustaceans and molluscs.

Modern shark groups, the neoselachians, arose in the Triassic, and radiated dramatically during the Jurassic and Cretaceous to reach their modern diversity of 35 families. The neoselachians have wider-opening mouths than earlier sharks, and they have particular adaptations for gouging masses of flesh from other large animals. The snout is usually longer than the lower jaw, so the mouth is set back beneath the front of the head. When the jaws open, they are thrust forwards, exposing rows of sharp triangular teeth, and the whole apparatus is ideal for devouring vast mouthfuls of flesh.

The body shape of neoselachians is even more bullet-like than in their ancestors, and the pectoral fins are wider and more flexible. Neoselachians range in size from common dogfishes (0.3–1m long) to basking and whale sharks (17m long), but these monsters are not predators: they feed on krill which they filter from the water. The first neoselachians were nearshore hunters that fed on bony fishes and squid, while later members of the group became faster offshore hunters.

The skates and rays, unusual neoselachians, are specialised for life on the sea floor, having flattened bodies, with the eyes on the top surface, and the mouth beneath. The pectoral fins are broad, and swimming movement is produced by sinuous up-and-down undulations of the pectoral fins.

Dapedium, a deep-bodied neopteygian bony fish from the Lower Jurassic. This was one of the major radiations of bony fishes that occurred in the Triassic and Jurassic.

New groups of bony fishes appeared in the Triassic and Jurassic, especially the basal neopterygians, such as *Dapedium*, a small deep-bodied form. It had lighter scales than those found in earlier bony fishes, and a jaw apparatus that could be partly protruded, hence providing a wider gape. *Semionotus*, another holostean, does not have such a deep body, and vast shoals have been found in the deep lakes of the Newark Supergroup, ranging in age from late Triassic to Early Jurassic (see pages 88–89). There are a few living relatives of the early bony fishes, the bichir, the sturgeon, the paddlefish, the gars, and the bowfin, but most modern fishes are teleosts. The teleosts arose in the Jurassic, and radiated mainly in the Cretaceous (see page 107).

The three-phase radiation of bony fishes, palaeonisciforms in the Permo-Carboniferous, basal neopterygians, and teleosts, is paralleled by the three-phase radiation of sharks. New adaptations in the bony fishes permitted them to swim faster, to diversify their feeding strategies, and hence to become more abundant. Likewise, new adaptations permitted the various shark groups to swim faster and perhaps to conquer new environments. It is impossible to say which set of evolutionary radiations came first: the bony fishes had to swim faster to escape their sharky predators, and the sharks had to swim faster to catch their bony fish prey.

Dragons of the Deeps

During the Jurassic, several reptile groups ruled the waves. Specialised crocodilians, the steneosaurids and metriorhynchids, became highly aquatically-adapted, having paddle-like limbs, and long thin snouts armed with interlocking sharp teeth, designed for trapping and retaining fish prey. The metriorhynchids in particular were fully committed to a life in the sea, having

Two ichthyosaurs from the Lower Jurassic of Holzmaden, south-west Germany. These dolphin-like reptiles fed on fishes and on swimming cephalopods, such as ammonites and belemnites.

streamlined bodies, completely enclosed paddles, and a long tail fin to assist swimming.

The ichthyosaurs ('fish reptiles') were fish-shaped animals, entirely adapted to life in the sea, but almost certainly evolved from land-living diapsids. Their Triassic forebears (see pages 62–63) had hit on the basic body design that was to be maintained by the group until its extinction in the Late Cretaceous. Ichthyosaurs had a long thin snout lined with sharp teeth, and they fed on ammonites, belemnites, and fishes. Exquisite preservation of many specimens, particularly from the Early Jurassic of southern Germany, shows the outlines of the dorsal fin and the paddles, and there is even preservation of the original skin. Ichthyosaurs swam by beating the body and tail from side to side, and they used the front paddles for steering.

Ichthyosaurs produced live young, as shown by numerous beautifully preserved specimens of females with enclosed embryos. There was a debate for a long time about whether these small ichthyosaur skeletons contained within the bodies of larger forms might have been eaten cannibalistically. Now, nearly 100 specimens of such large and small ichthyosaur associations have been collected. They come mainly from the Lias (Lower Jurassic) of Holzmaden, a series of quarries near Stuttgart in south-west Germany, and a few have also been found in the Lias of southern England. Recent studies have suggested that these are in fact embryos. The included skeletons are always the same species as the adult, and the small skeletons never show any signs of breakage, dismemberment, or dissolution of bone material. When an animal is eaten by a predator, it is usually broken up in the mouth, and the skeleton then is broken and dissolved by stomach activity. This is not seen in the case of ichthyosaurs. What is seen is that the stomach region, located in a different position from the 'embryos' may contain a mass of horny hooklets from the tentacles of belemnites, or debris of fish scales.

The included 'embryos' in large ichthyosaurs are nearly always arranged in a particular way. The smallest ones are coiled up in a classic 'embryo' posture, but beyond a certain size, they straighten out, and they are nearly always found with their heads pointing forwards with respect to the putative mother. Typically, the adults contain one or two small skeletons: one specimen had eleven inside. The head-forwards posture is interpreted as a factor in ensuring that the young were born tail-first. This is an adaptation seen in whales today. The juvenile slips out of the birth canal, and only at the last minute its head emerges, it severs the umbilical cord, and it swims to the surface immediately to take its first breath. If baby whales (and ichthyosaurs) were born head-first, they would drown.

Brachiosaurus, one of the largest dinosaurs, a sauropod from the Late Jurassic of Tanzania and North Africa (below). *Brachiosaurus was nearly 23m long, and it was especially tall, at 12m, because of its very long forelimbs. A spectacular complete skeleton was collected early in the 20th century, and is now on show in Berlin, Germany* (see pages 98–99).

The second major marine reptile group were the plesiosaurs ('distant reptiles'). Most plesiosauroids had long necks and small heads, but the pliosauroids were larger, and had short necks and large heads. Plesiosauroids fed mainly on fishes and ammonites, using the long neck like a snake to dart after fast-moving prey, and they swam by beating their paddles in a kind of 'flying' motion. Pliosauroids probably also fed on fishes and ammonites, but the larger species became top predators, preying on ichthyosaurs and plesiosauroids. Both plesiosauroids and pliosauroids became important in the Early Jurassic, and they radiated through the Jurassic and Cretaceous. They arose from Triassic ancestors similar to the nothosaurs (see page 62).

Jurassic Dinosaurs

Early Jurassic dinosaur faunas (see pages 90–91) were very similar to those of the Late Triassic. They were dominated by prosauropods as herbivores, and ceratosaurs as carnivores. Towards the end of the Early Jurassic, the first sauropods, descended from the prosauropods, arrived on the scene. These early sauropods, represented by isolated finds, *Vulcanodon* from South Africa, *Barapasaurus* from India, and isolated remains from Germany and elsewhere, showed key changes in skull pattern (larger eyes, nostrils shifted back, teeth restricted to the front of the jaw) and in the skeleton (more vertebrae in the neck, reinforced limb girdles, pillar-like limbs, simplified foot skeletons). This was the beginning of the rise of the monster dinosaurs (see pages 94–99).

The sauropods were mainly large and very large animals, some of them, such as *Brachiosaurus* reaching lengths of 23m or more, and heights of 12m. The sauropods were most abundant during the Jurassic, and they became much less significant during the Cretaceous, although some sauropod groups survived until the very end of that period. These giant dinosaurs have posed fascinating problems in palaeobiological interpretation. When the monster sauropods of the Late Jurassic were first discovered in the American Midwest in the late 19th century (see pages 100–101), many palaeontologists thought that they were too big to have lived fully on land. It was assumed that the sauropods lived in lakes, supporting their bulk in the water, and feeding on waterside plants. New evidence shows, however, that life on land was quite possible, and indeed the long neck of *Brachiosaurus* made it like a super-giraffe, a dinosaur that could feed on leaves from very tall trees, well out of the range of any other animal.

More recently, some palaeontologists have suggested that sauropods might have been able to gallop at speed, and hoist themselves up on their hind legs to reach even higher in the trees. These ideas are not likely since both activities would probably have broken the animal's legs. Experiments with modern bones show how much strain they can withstand. It is a simple matter to work out the breaking point of ancient bones, based on their diameter and their density, and these experimental studies show that a galloping *Brachiosaurus* would have collapsed in a pile of shattered limbs.

The other saurischian group, the Theropoda, radiated during the Jurassic. The ceratosaurs continued from ancestors like *Coelophysis* from the Late Triassic (see pages 74–75) to radiate into larger forms in the Jurassic. Some, like *Dilophosaurus* from the Early Jurassic of North America, and *Ceratosaurus* from the Late Jurassic of Tanzania and North America (see pages 98–99, 102–103) had cranial excrescences of one sort or another: parallel plate-like crests in the former, low horns in the latter. These may have been used by the males in sexual displays to attract a mate.

A new theropod group, the tetanurans, rose to prominence in the Jurassic. These were mainly larger theropods, animals such as *Megalosaurus* and *Gasosaurus* from the Middle Jurassic (see pages 92–95) and *Allosaurus* from the Late Jurassic (see pages 102–103). The tetanurans had a distinguiished evolutionary record, giving rise also to the birds in the Jurassic (see pages 86–87), and an astonishing array of theropods small and large in the Cretaceous (see pages 111–112).

The sauropods and theropods are, of course, saurischian dinosaurs (see pages 72–73). The other major dinosaurian subgroup, the Ornithischia, includes only herbivores, and these are divided into five or six distinctive groups, three of which radiated in the Jurassic, the stegosaurs, ankylosaurs, and ornithopods.

Stegosaurs and ankylosaurs form together the Thyreophora. The first thyreophorans appeared in the Early Jurassic, animals like *Scelidosaurus* in England and *Scutellosaurus* in North America. The group split in the Middle Jurassic into the stegosaurs and ankylosaurs, but the latter group is known in the Jurassic only from limited remains, and it radiated mainly in the Late Cretaceous (see pages 130–133). The stegosaurs, on the other hand, are best-represented in the Middle Jurassic of China (see pages 94–95) and the Late Jurassic of Tanzania and the American Midwest (see pages 98–103). *Stegosaurus*, from the Late Jurassic of North America, had a row of bony plates along the middle of its back, which may have aided temperature-control.

The ornithopod dinosaurs are known first in the Late Triassic and Early Jurassic from a number of small rare forms (see pages 90–91). By Late Jurassic times, ornithopods such as *Camptosaurus* had achieved moderately large size, and they were coming into their own in a landscape still dominated by sauropods. The sauropods declined after the end of the Jurassic, and the ornithopods became dominant as herbivores in the Early Cretaceous. They owed their success, in all probability, to their massive batteries of teeth, which allowed effective browsing on tough vegetation. In addition, they were all bipeds, and they specialised in speed of movement to escape from larger predators. Close relatives of the ornithopods, the ceratopsians and pachycephalosaurs, arose in the Cretaceous.

Pterosaurs

The pterosaurs were proficient flapping flyers which filled the Jurassic skies. The group had arisen in the Late Triassic, and it seems that pterosaurs are close relatives of the dinosaurs. They are characterised by a lightweight body, narrow hatchet-shaped skull, and a long narrow wing supported on a spectacularly elongated fourth finger of the hand. The bones of the arm and finger support a tough flexible membrane that could fold away when the animal was at rest, and stretch out for flight. Pterosaurs were covered with hair, which can be seen in some of the best-preserved specimens, especially from the Late Jurassic of Germany and Kazakhstan, and they were almost certainly warm-blooded.

Pterosaurs of the Jurassic include a variety of basal forms, the rhamphorhynchoids, such as *Rhamphorhynchus*, and the pterodactyloids, such as *Pterodactylus*, both from the Late Jurassic of Germany. The rhamphorhynchoids retained the primitive long tail, and never exceed the size of a seagull. Pterodactyloids had short tails, and some of them, in the Late Cretaceous, achieved gigantic size, with wingspans of 12m or more (see page 109).

There has been considerable debate about the modes of locomotion of the pterosaurs. The first set of debates has concerned their ability to fly, whether they merely glided on their bat-like wings, or flapped actively. Most palaeontologists accept that pterosaurs were active flapping flyers, but their flight dynamics are disputed. Some reconstruct pterosaurs all with long slender wings, shaped like the wings of a gull. Others argue that the wing membrane not only attached along the arm and elongate fourth finger, but that it also ran back along the side of the body and on to the hindlimb. A slender gull-like wing would be an adaptation for long-sustained flight, while a broad wing would be more manoeuvrable.

Pteranodon, one of the largest pterosaurs, from the Late Cretaceous of North America. Pterosaurs were efficient flapping fliers which had evolved their flight capability independently of the birds.

The other debate concerns the locomotion of pterosaurs on land: did they run about readily, or were they awkward waddlers? According to the first view, pterosaurs had slender limbs held in an upright posture, just as in bipedal dinosaurs, and they could run rapidly around, with their wings neatly tucked back like a pair of smartly folded umbrellas. The other view is that pterosaurs had broadly sprawling hindlimbs, and that they tumbled about on land, like broken kites, using their wings as additional props.

Origins of the Modern Groups

Modern amphibians, frogs and salamanders, are a minor group, consisting of 4000 species that lived generally in damp conditions. The oldest frog specimens date from the Early Triassic, and these suggest that the modern amphibians arose from small temnospondyls of the Palaeozoic (see pages 46, 52–53). Several fossil frog groups are known from the Jurassic, including some modern lines. These show specialisations for jumping: the hindlimbs are long and the hip bones are reinforced to withstand the impact of landing. The head is broad, the jaws are lined with small teeth, and most frogs feed by flicking out a long sticky tongue out to trap insects.

The salamanders and newts date from the Jurassic, and consist of modest-sized long-bodied swimming predators. Isolated salamander fossils are known from the Middle Jurassic of England, but the first well-preserved form is *Karaurus* from the Late Jurassic of Kazakhstan. It has a broad skull with a neat semicircular front margin, and the jaws are lined with numerous small peg-like teeth.

The crocodiles, again a modest group today, consisting of about 15 species, arose in the Late Triassic (see page 62). The first crocodilians were light-weight bipeds that fed on insects. The Early Jurassic crocodilian *Protosuchus* looked more like a modern crocodile, but it was still largely terrestrial in habits. It walked on all fours, and had an extensive armour of bony plates. Crocodilians were more diverse and abundant during the Jurassic and Cretaceous than they are now, and they included a variety of marine, fresh-water, and terrestrial forms.

The squamates, lizards and snakes, arose in the Late Triassic. The key forms then were sphenodontids, snub-nosed lizard-sized animals that fed on plants and insects. The group dwindled, however, during the Jurassic and Cretaceous, and there is only one living representative, *Sphenodon*, the tuatara of New Zealand, a famous 'living fossil'. The sphenodontids are close relatives of the lizards.

Archaeopteryx, *the oldest bird* (right), *known from seven skeletons from limestone quarries at Solnhofen, Bavaria, southern Germany. The skeletons show a mix of dinosaur-like features—a long bony tail, three-fingered hands with claws, and jaws with teeth. But* Archaeopteryx *was a bird: it had feathers and it could fly.*

The first true lizards are known from the Mid and Late Jurassic, mostly represented by isolated partial skulls, but with one or two well-preserved fossils from those early times. These early lizards show the characteristic mobility of the skull: the bar beneath the lower temporal opening is incomplete, the quadrate is mobile, and the snout portion of the skull can tilt up and down. This unusual flexibility of the skull allows lizards to open their mouths wider than would otherwise be possible, in order to swallow large prey, and it helps also in probing and burrowing. The process of loosening of the skull was taken even further in the snakes, a group known first in the Early Cretaceous. Snakes have such mobile skulls that they can virtually disarticulate their jaws and major skull bones in order to swallow prey animals that are several times the diameter of the head.

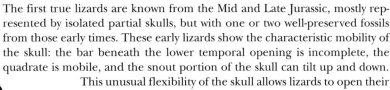

Megazostrodon, *one the first mammals from the Early Jurassic of South Africa. This small insectivorous nocturnal animal was typical of the Mesozoic mammals, most of which remained small. Mammals became larger only after the extinction of the dinosaurs.*

One of the most famous fossils is *Archaeopteryx*, the oldest known bird. The first specimen was found in Upper Jurassic sediments in Bavaria, southern Germany, in 1861, and was hailed as the ideal 'missing link' or proof of evolution in action. Here was an animal with a beak, wings, and feathers, so it was clearly a bird, but it still had a reptilian bony tail, claws on the hand, and teeth. Since 1861, six more skeletons have come to light, the last two in 1987 and 1992.

Archaeopteryx was about the size of a magpie, and it fed on insects. The claws on its feet and hands suggest that *Archaeopteryx* could climb trees, and the wings are clearly those of an active flying animal. This bird could fly as well as most modern birds, and flying allowed it to catch prey that were not available to land-living relatives.

The origin of birds has been debated for a long time. Thomas Henry Huxley realised the truth of the matter, soon after the discovery of *Archaeopteryx*, when he showed how similar its skeleton was to that of certain dinosaurs. However, for the next hundred years, many ideas were suggested, and Huxley's view was largely forgotten. Most popular was the idea that birds originated directly from basal archosaurs of the Triassic, but others have suggested an origin from Late Triassic crocodilian relatives, or from a joint ancestor with mammals. The skeleton of *Archaeopteryx*, however, is very like that of *Deinonychus*, especially in the details of the arm and hind limb, showing that birds are small flying tetanuran theropod dinosaurs. Birds remained rare until the Late Cretaceous when new marine forms appeared, as well as some of the modern groups, although most of those arose only during the past 65 million years of the Cenozoic.

The first mammals, small insect-eaters in the Late Triassic and Early Jurassic, probably hunted at night. Mammals remained small through most of the Jurassic and Cretaceous, and they failed to make an impact as long as the dinosaurs existed. Several lines of insectivorous, carnivorous, and herbivorous forms appeared, some of them adapted to climbing trees, but only one or two of these groups survived the mass extinction event 65 million years ago. These included the ancestors of the three living mammal groups, the monotremes, marsupials, and placentals, which had arisen in the Cretaceous.

The Jurassic was a time of major change in life of the sea and land. Major groups of reptiles rose to dominate on land, in the air, and in the sea. At the same time, the origins of many thoroughly modern animals—frogs, lizards, birds, mammals—can also be traced to these times.

The Jurassic World

The Jurassic period was warm and humid, and ceratosaurs, tetanurans, stegosaurs, and sauropods flourished.

The Late Triassic was characterised, in many parts of the world, by hot arid conditions (see pages 68–69), but climates became more humid in the Jurassic. This climatic change may be associated with major plate tectonic movements, and with global changes in sea level. There were also major continental movements. The supercontinent Pangea had been at its maximum point of fusion in the Late Triassic, about 230 Myr ago, when all land masses were joined. Rifting began in the Late Triassic between North America, Africa, and Europe. Narrow seas opened up, and the Earth's crust tore apart in a series of huge faults that ran roughly south-west to north-east. As the faults opened up, marine waters flooded in periodically from east and west, and then evaporated, leaving salt deposits. The initial rifting and partial flooding phase lasted for 20–30 million years in the Late Triassic and Early Jurassic, but by the Middle Jurassic, the zone had widened sufficiently to contain a permanent belt of sea: the embryo North Atlantic.

The initial Atlantic fault zones are very clear particularly all the way along the eastern seaboard of North America, where they formed a series of lakes, in many ways like the rift valley lakes in east Africa today. The American rift lakes filled with great thicknesses of freshwater sediments, termed the Newark Supergroup, and it has been possible to measure many kilometres thickness of Newark sediments, and to record various scales of cyclicity in patterns of deposition. At the scale of millimetres and centimetres, the Newark lake deposits seem to show annual varves,

Semionotus (above), a fish from the Late Triassic and Early Jurassic rift lakes of eastern North America.

Jurassic dinosaurs included long-necked plant-eating sauropods such as Diplodocus (far right) and the plate-backed stegosaurs such as Stegosaurus (below). The smaller herbivores were preyed on by carnivores such as Megalosaurus (bottom), but the sauropods (opposite) were probably so large that they escaped predation.

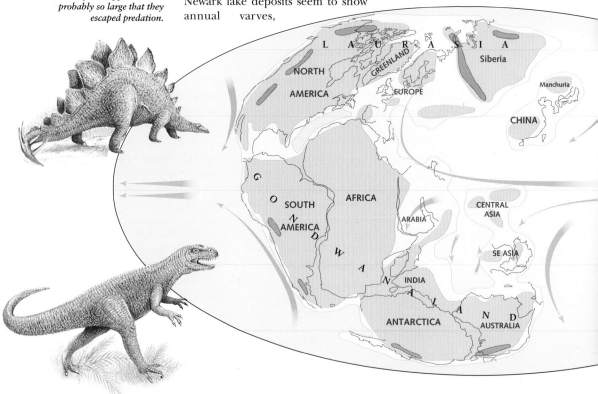

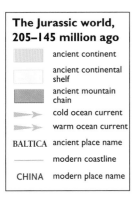

The world in the Middle and Late Jurassic (below) shows the beginnings of the modern distribution of the continents. The oceans began to split along modern lines. The North Atlantic was the first to open, as a narrow belt running SW to NE, as North America broke away from Africa. Major rises in sea level also flooded large parts of Europe and central Asia, causing further fragmentation land masses. Jurassic dinosaurs evolved on all land masses and, though separated by oceans at times of maximum flooding, phases of regression of the sea, and allowed the dinosaurs to move relatively freely around the world.

The Jurassic world, 205–145 million ago

	ancient continent
	ancient continental shelf
	ancient mountain chain
→→	cold ocean current
→→	warm ocean current
BALTICA	ancient place name
··········	modern coastline
CHINA	modern place name

many millions of them, reflecting cycles of dry and wet periods, times of rapid erosion and minimal erosion, times of high oxygenation of the waters, and poor oxygenation. The annual varves are then grouped in longer-term cycles on scales from 20,000 to 400,000 years. These so-called Milankovitch cycles record astronomical fluxes which can be observed today.

The string of Newark lakes varied greatly in pattern and size during their 50 million year history. As Atlantic opening proceeded, the rift valley topography changed, and lakes amalgamated and divided. They were populated by abundant fishes and molluscs, and the shorelines were rich in plant and animal life. Dinosaurs left tracks, as they came down to drink and feed. The fishes, particularly the heavy-scaled semionotids, lived in huge shoals, and in places today single bedding planes in the lake sediments can yield thousands of specimens. Each lake seemed to have its own populations of semionotids, and as many as 25 species together, each presumably specialising in subtly different food supplies. This proliferation of fish populations has been interpreted as a species flock, evidence of dramatic evolutionary divergence, as seen today among cichlid fishes in the African rift valley lakes.

As the North Atlantic opened, forming a division initially between the land mass consisting of North America and Europe in the north and Africa and South America in the south, eastern parts of Pangea also began to break free. Parts of Siberia, China, and south-east Asia separated as island land masses. These divisions were related partly to rifting and continental fission, but also to major sea level change. Compared to the Triassic, sea levels were dramatically elevated during the Jurassic, and large areas of former lowlands were flooded. This is seen especially dramatically in Europe. The lands of Britain and central Europe that were tramped by the giant plateosaurs and other dinosaurs in the Late Triassic (see pages 76–77) were speedily flooded during the Rhaetic transgression which extended over much of western Europe about 205 Myr ago, at the very end of the Triassic.

This transgression is seen very clearly as one follows sequences of Triassic and Jurassic rocks in Britain and Germany. Hundreds of metres of classic red-coloured continental sediments of the Triassic, containing the bones and footprints of dinosaurs and other evidence of dry conditions, are cut across suddenly by grey and black marine sediments of the Rhaetic. These contain sea shells and the bones of marine reptiles, and they herald the flooding of Europe which lasted for much of the Jurassic. The seas withdrew from north-west Europe in the Middle Jurassic, and marginal lowland deposits accumulated in some areas, and these have yielded important dinosaur faunas (pages 92–93). A second phase of flooding occurred in the Late Jurassic, and seas flooded NW Europe, a marine phase that came to an end at the transition from the Jurassic to Cretaceous.

These dramatic global changes in sea level, or eustatic sea level changes, are probably related to the increased rate of rifting. As new oceans opened up, oceanic crust was created by mid-ocean ridges (see pages 50–51), and these active volcanic zones raised the level of the deep ocean floor.

Dinosaurs of the Early Jurassic

New groups of dinosaurs appeared in the Early Jurassic, but evidence is patchy since terrestrial deposits are limited, and dinosaur remains are often rare.

The Early Jurassic was an important phase in dinosaurian evolution, marking a transition from Triassic-style faunas, with prosauropods such as *Plateosaurus* and small ceratosaur predators like *Coelophysis* (see pages 74–77) to Jurassic-style faunas with large meat-eaters, the first true sauropods, and armoured dinosaurs. In Europe, particularly, thanks to the Early Jurassic marine transgression (see pages 88–89), dinosaur bones are rare isolated elements that were washed into the shallow seas. In North America, Early Jurassic dinosaur remains are also sporadic and dinosaur footprints are rather commoner than skeletons.

The best dinosaur faunas of the Early Jurassic are known from China. In the Lufeng basin, in Yunnan province in southern China, continental sediments 1 kilometre thick accumulated. The commonest dinosaur in these sediments is the prosauropod *Lufengosaurus*, similar to *Plateosaurus* (see pages 76–77). *Lufengosaurus* was 6 metres long, and stood 3 metres high when it reared up to gather leaves from trees using its strong arms and fingers.

Other dinosaurs from the Lufeng Formation include the small carnivore *Lukousaurus*, which probably hunted small non-dinosaurs. There was also a new form, *Tatisaurus*, a primitive ornithischian dinosaur. This little biped fed on plants which it cut up with strong little teeth. *Tatisaurus*, and relatives from elsewhere, mark the beginning of the radiation of ornithischians, slow in the Jurassic, but much more dramatic in the Cretaceous (see pages 118–119).

Other Lufeng animals that ran about at the feet of the dinosaurs, were the mammal-like reptiles *Bienotherium* and *Yunnanodon*, broad-toothed herbivores belonging to a surviving group of cynodonts, the tritylodonts, which lived on through the Middle Jurassic. In addition, there was a small mammal, *Sinoconodon*, no larger than a rat, and probably living a secretive nocturnal existence, feeding on insects and worms.

The Early Jurassic Stormberg Group of South Africa has also been an important source of Early Jurassic dinosaurs. For years, these red-coloured continental sediments had been dated as Late Triassic in age (as indeed had the Purple Beds and Red Beds of the Lufeng Basin, and the bulk of the Newark Supergroup (see pages 88–89). Re-assessment of the ages of these units in the 1980s caused a major rethink. Stormberg dinosaurs include prosauropods and small ornithischians. The oldest sauropod *Vulcanodon* comes from similar-age units in Zimbabwe.

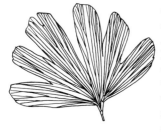

Jurassic plants were mainly of primitive types, gymnosperms such as Sagenopteris. *Gymnosperms bore seeds, and modern groups include conifers, cycads, and ginkgos. Other Jurassic plants included ferns, horsetails, and lycopods, small tree-like plants. These groups all survive today, even if in some cases at low diversities, but their coarse vegetation formed the bulk of the diet of herbivorous dinosaurs.*

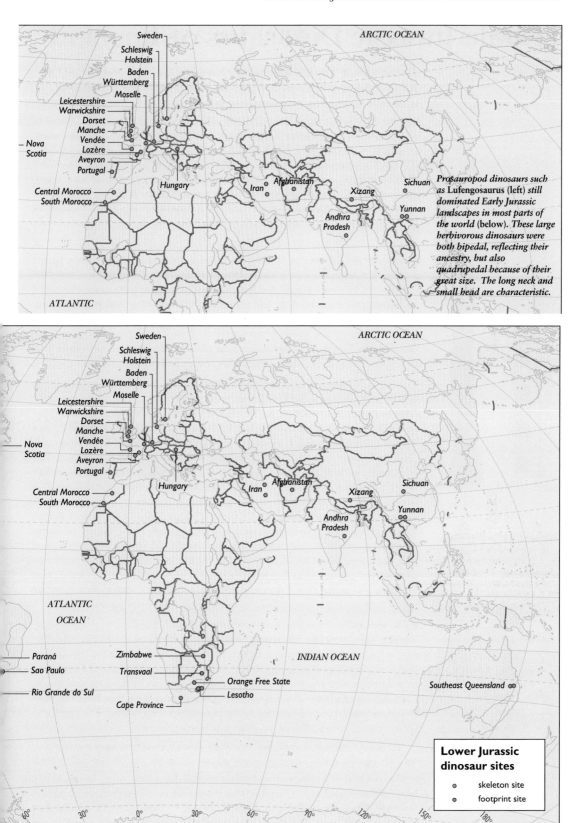

Sweden
Schleswig Holstein
Baden
Württemberg
Moselle
Leicestershire
Warwickshire
Dorset
Manche
Vendée
Lozère
Aveyron
Portugal
Nova Scotia
Hungary
Central Morocco
South Morocco
Iran
Afghanistan
Xizang
Sichuan
Andhra Pradesh
Yunnan

ARCTIC OCEAN

ATLANTIC

Prosauropod dinosaurs such as Lufengosaurus *(left) still dominated Early Jurassic landscapes in most parts of the world (below). These large herbivorous dinosaurs were both bipedal, reflecting their ancestry, but also quadrupedal because of their great size. The long neck and small head are characteristic.*

Sweden
Schleswig Holstein
Baden
Württemberg
Moselle
Leicestershire
Warwickshire
Dorset
Manche
Vendée
Lozère
Aveyron
Portugal
Nova Scotia
Hungary
Central Morocco
South Morocco
Iran
Afghanistan
Xizang
Sichuan
Andhra Pradesh
Yunnan

ARCTIC OCEAN

ATLANTIC OCEAN

Paraná
Sao Paulo
Rio Grande do Sul

Zimbabwe
Transvaal
Cape Province
Orange Free State
Lesotho

INDIAN OCEAN

Southeast Queensland

Lower Jurassic dinosaur sites

- skeleton site
- footprint site

60° 30° 0° 30° 60° 90° 120° 150° 180°

91

Middle Jurassic Dinosaurs of England

In an interlude between the marine episodes of the Early and Late Jurassic, parts of England and Scotland emerged from the seas, and dinosaurs lived on the transient land masses.

The first recorded dinosaur discovery, the end of a limb bone of a carnivorous dinosaur reported by Robert Plot in 1676 (see pages 22–23), was made in Middle Jurassic rocks in Oxfordshire. Also, more substantial remains of the same dinosaur were found nearby at Stonesfield in 1818, and they were named *Megalosaurus* in 1824, the first dinosaur in the world to be named. Since then, the Cotswolds region of southern England, in the counties of Oxfordshire and Gloucestershire, has yielded a constant supply of remains of dinosaurs and other fossil vertebrates of Middle Jurassic age.

A typical site was investigated in huge detail at Hornsleasow in Gloucestershire. Here, in 1977 Kevin Gardner, a keen fossil collector, found a large *Megalosaurus* tooth. Geologists at Gloucester City Museum sampled the site, finding many more teeth and bones. A team of enthusiastic amateur archaeologists excavated 20 tonnes of fossiliferous clay, and bagged it up. They also found large bones of the sauropod dinosaur *Cetiosaurus*, lying in the quarry floor. The project transferred to Bristol University, where more of the clay was sieved using an automated process. By 1994, all 20 tonnes had been processed, and 15,000 identifiable bones had been extracted.

The site turned out to be a small pond that had trapped the bones of numerous animals. Perhaps the *Cetiosaurus* became trapped in the mud, and some *Megalosaurus* came to scavenge. The majority of the bones came from smaller animals. There were numerous fish scales and teeth—both bony fishes and sharks—which probably lived in the pond. There were also some crocodile teeth and bones, some turtle shell fragments, and some very rare salamander fragments, some of the oldest in the world. These too probably inhabited the pond. The other bones and teeth represented animals that probably lived on shore, but might have been drowned in the pond while drinking, or they might have been pulled in by the crocodiles, or their bones might simply have been carried by streams draining into the pond. These terrestrial forms included a variety of other dinosaurs, theropods, sauropods, and stegosaurs, rare lizards (again, some of the oldest in the world), mammals, and mammal-like reptiles (some of the last in the world). This is probably one of the most intensively sampled single sites and an extraordinarily detailed picture of the life in and around a single Middle Jurassic pond has been reconstructed.

Palaeogeographic reconstructions of the British Isles during the Middle Jurassic show how the dinosaurs became established in the Cotswolds. The seas, which had flooded much of Britain during the Early Jurassic, withdrew, and low-lying coastal lands appeared between the Anglo Brabant Land mass near London and the Welsh and Pennine landmasses to the west and north. Dinosaurs spread over this region, living on swampy areas that built out over the limestones that had lain below the sea a short time before. The setting in Oxfordshire in the Middle Jurassic has been compared to the Everglades swamps of Florida, where alligators live in profusion today.

The Middle Jurassic lands extended also in north-west Scotland, and similar sediments had been known on the Isle of Skye for a long time. In the 1970s, some isolated bones of crocodiles and mammals were found there, as well as a single dinosaur footprint. Finally, in 1992, 1993, and 1995, the first dinosaur bones were found, at three separate localities on Skye.

For years, Scotland has sought its own dinosaur. A flurry of excitement greeted the announcement, in 1910, of a dinosaur, Saltopus, *from the Late Triassic of Elgin, in north-east Scotland. Close study since then has not confirmed that this is definitely a dinosaur: it might be, but the remains are not good enough to be sure. Then a single dinosaur footprint was found on Skye in the 1970s. Did the dinosaur hop in, and then disappear? Seemingly not. In 1992, 1993, and 1995, three definite dinosaur bones were found on Skye, including the small limb bone of a theropod like* Dilophosaurus (above), *one of the characteristic dinosaurs of the English Middle Jurassic. The other Scottish bones are from a smaller meat-eater.*

Bathonian outcrop

Worldwide, Middle Jurassic dinosaurs are not very well known. Major localities for dinosaurs of this age are found in Britain and China, but almost nothing is known from other parts of the world. In Britain, Middle Jurassic dinosaurs have been found especially in the Bathonian (above), *which was deposited in coastal areas* (right).

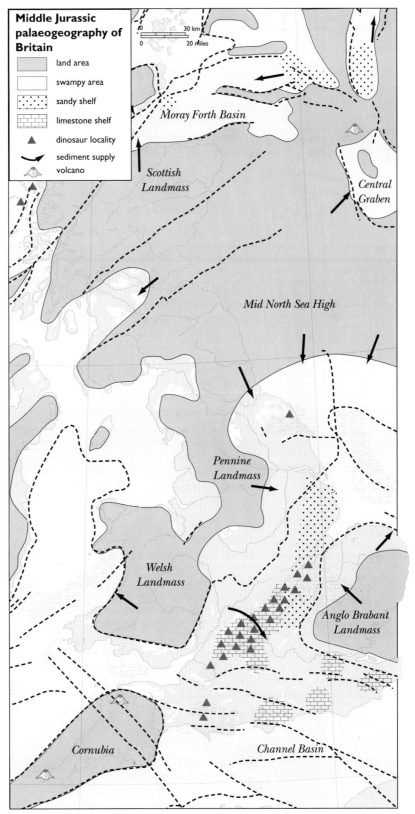

Middle Jurassic palaeogeography of Britain

- land area
- swampy area
- sandy shelf
- limestone shelf
- ▲ dinosaur locality
- ↜ sediment supply
- 🌋 volcano

Moray Forth Basin

Scottish Landmass

Central Graben

Mid North Sea High

Pennine Landmass

Welsh Landmass

Anglo Brabant Landmass

Cornubia

Channel Basin

Middle Jurassic Dinosaurs of China

Collecting in the past 20 years has produced a staggering number of new dinosaur discoveries from the Middle Jurassic of China.

For centuries, dinosaurs bones had been dug up in China and used as medicine. But to westerners China remained a mysterious land, and it lacked indigenous palaeontologists. In 1927 a Sino-Swedish expedition travelled through the northern provinces —Inner Mongolia, Ningxia, Gansu, and Xinjiang—where dinosaur fossils were collected mainly from Upper Jurassic and Cretaceous deposits. The resulting publications attracted world attention to China, but the Second World War prevented further substantial activity until the 1950s.

The skull of Huayangosaurus *(above), the oldest-known stegosaur. The skull is low and tubular in shape, and, like all later stegosaurs, Huayangosaurus was not blessed with outstanding brain power. The long narrow snout housed large nasal organs, so this dinosaur had a good sense of smell. The teeth are small leaf-shaped structures, useful for piercing and cutting relatively soft vegetation, but no use for dealing with woody plant tissues.*

A striking discovery was made in 1977 at Dashanpu in Shandong Province in rocks of Middle Jurassic age. The Dashanpu dinosaur quarry was opened up, and over 100 complete dinosaur skeletons excavated, together with fishes, amphibians, pterosaurs, and other reptiles. Dashanpu has yielded eight dinosaur species. The commonest forms, representing 94% of the finds, are the sauropods *Shunosaurus*, *Omeisaurus*, and *Datousaurus*. Other herbivores include the small ornithopod *Xiaosaurus*, and the stegosaur *Huayangosaurus*, and these were preyed on by the theropods *Gasosaurus* and *Xuanhanosaurus*.

Shunosaurus was the first dinosaur to be found at Dashanpu. It was a modest-sized sauropod, 20 metres long, and broadly similar to *Cetiosaurus* from the Middle Jurassic of England (see pages 92–93). In detail, the skull of *Shunosaurus* shows its importance for an understanding of sauropod evolution. The nostrils are set a little way back from the tip of the snout, and the eye socket has moved back a little, but the row of blunt teeth extends back to the level of the eye socket. These are all primitive characteristics when compared with the Late Jurassic sauropods (see pages 98–103), in which the nostrils are right over the top of the skull, the eye socket has moved well back, and the teeth occupy only the front part of the snout. *Shunosaurus* is classed in a basal sauropod family, the Euhelopodidae.

Huayangosaurus, one of the most celebrated dinosaurs from the Middle Jurassic sediments of Dashanpu Quarry, is known from nearly complete skeletons (below). The spines and plates along the back show that this was a stegosaur, and their shape suggests that they may have been largely defensive structures. The hindlimbs are much longer than the forelimbs, and this confirms that the armoured dinosaurs had a bipedal ancestor in the Early Jurassic or Late Triassic. Huayangosaurus was about 4 metres long.

The other two sauropods from Dashanpu are probably closer relatives of *Cetiosaurus* from the Middle Jurassic of England, and they are somewhat more advanced than *Shunosaurus*. The second sauropod, *Datousaurus*, is estimated to have been perhaps 15 metres long, but the skeleton is incomplete. The largest sauropod from Dashanpu, *Omeisaurus*, is represented by a spectacular complete skeleton, fully 20 metres long. *Omeisaurus* had 16 vertebrae in its enormously elongated neck, an advance over the 12–13 neck vertebrae of *Shunosaurus* and *Datousaurus*. The other herbivores include the hypsilophodontid *Xiaosaurus*, a relative of the Early Cretaceous *Hypsilophodon* from England (see pages 118–119). *Xiaosaurus* was the smallest dinosaur

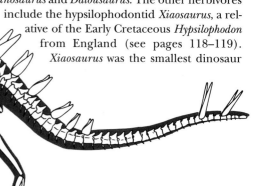

The Middle Jurassic fauna from Sichuan Province included dinosaurs large and small (above). Middle Jurassic and Late Jurassic dinosaur faunas have been found in several regions of China (below). The first scientific dinosaur excavations in China were carried out by the Sino-Swedish Expedition of 1927 which uncovered dinosaurs of Late Jurassic and Cretaceous age.

in the fauna, reaching only 1.5 metres in length. It was a lightly-built biped that had long powerful legs, and was clearly fast and agile, a useful defence against the predatory dinosaurs.

The stegosaur *Huayangosaurus* protected itself in a different way, by the elaboration of its armour. *Huayangosaurus* is one of the most important finds from Dashanpu, since it is the oldest adequately known stegosaur. It was not the first armoured dinosaur—*Scelidosaurus* and *Scutellosaurus* from the Early Jurassic of England and the USA respectively—were ancestral armoured forms, but *Huayangosaurus* was the first stegosaur. In general form, *Huayangosaurus* is reminiscent of its much more famous descendant, *Stegosaurus* (see page 102), but it is older, and more primitive in many features. The skull is larger, and the spines and plates are not so broad. *Huayangosaurus* was 4 metres long, and its spikes were probably used largely for defence.

The predatory dinosaurs are rarer than the herbivores, and they are represented mainly by teeth. The large theropod *Xuanhanosaurus* was 6 metres long, but it is represented by incomplete remains. *Gasosaurus* is better known from a complete 4-metre long skeleton. *Gasosaurus*, probably a relative of *Megalosaurus* from the Middle Jurassic of England (see pages 92–93), has a powerful skull and strong three-fingered hands for grappling with prey.

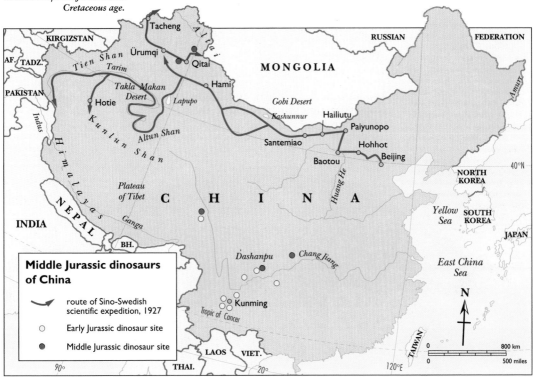

Middle Jurassic dinosaurs of China

⌒ route of Sino-Swedish scientific expedition, 1927

○ Early Jurassic dinosaur site

● Middle Jurassic dinosaur site

Dinosaurs of Gondwanaland

The dinosaurs of the southern continents are important in trying to disentangle the break-up of the supercontinent Gondwanaland.

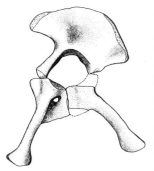

Pelvis of Patagosaurus *(above), one of the sauropods from the Jurassic of Argentina, reported in 1979 by José Bonaparte. This stocky sauropod had a shorter neck than the typical Late Jurassic sauropods, and it shows most similarity to* Cetiosaurus *from the Middle Jurassic of England.* Patagosaurus *is part of a dinosaur fauna that included another sauropod and an allosaurid theropod.*

The sauropods were the dominant plant-eaters of the Middle and Late Jurassic (below). Their massive size rendered them invulnerable to attack by the meat-eaters of their day. They fed on low plants, and probably used their long necks to forage over a great deal of ground, without having to move the rest of their bodies. Some palaeontologists believe that the sauropods could also rear up on their hind legs to feed high in trees, but that form of behaviour is uncertain in such massive animals.

Dinosaurs have been found in Middle and Late Jurassic rocks of South America, Africa, India, Australia, and Antarctica. During this time, these continents were still largely united as the supercontinent Gondwanaland, which had existed for some 200 million years or more (see pages 50–51). The process of fragmentation had, however, begun, with rifting in the embryonic South Atlantic, separating South America and Africa, and a drift eastwards of India, Antarctica, and Australia.

Dinosaurs were found rather late in the southern hemisphere, and the first finds were made by explorers from Europe. The first records came in the 1850s and 1860s, when dinosaur bones were sent back to England from South Africa. These specimens, from the Early Jurassic red beds of the Karoo Basin were from large prosauropod dinosaurs (see pages 76–77, 90–91), and they were named *Massospondylus* and *Euskelosaurus*. No further dinosaurs were reported from Africa until the 20th century. Robert Broom (1866–1951), a Scottish physician who had settled in South Africa (see page 54), reported on Cretaceous dinosaurs in 1904, and Early Jurassic dinosaurs in 1911. Many more dinosaurs were discovered in southern Africa after Broom's time, but most of these were from the Early Jurassic. The most important discoveries of Late Jurassic dinosaurs were made in 1907 in Tanzania, at Tendaguru (see pages 98–99).

Rich finds of Jurassic dinosaurs were also coming from South America at the same time. The first records from that continent were Upper Cretaceous dinosaurs from Argentina (see pages 124–125), reported in 1893. In the early part of the 20th century, many more Cretaceous dinosaurs were discovered in Argentina, and Friedrich von Huene (see pages 76–77) also found Triassic dinosaurs in Brazil in the 1920s, and later finds in Argentina and Brazil have turned out to include some of the first dinosaurs (see pages 70–71). Some dramatic discoveries of Jurassic dinosaurs were made in Argentina in the 1970s, by José Bonaparte, a major figure in palaeontology in that country, and by some of his students.

The Chubut locality in Patagonia, dated as Middle to Late Jurassic, produced a fauna consisting of some theropods and sauropods. The theropod,

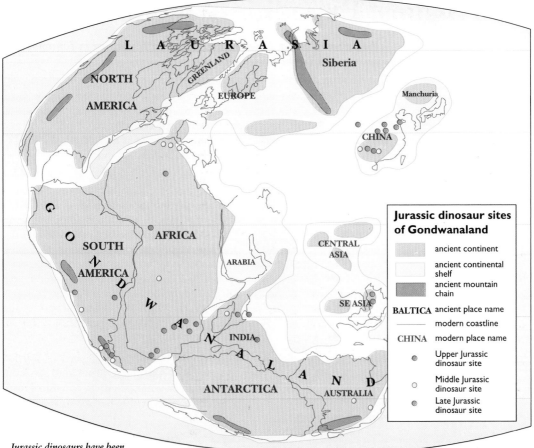

Jurassic dinosaur sites of Gondwanaland

- ancient continent
- ancient continental shelf
- ancient mountain chain
- **BALTICA** ancient place name
- —— modern coastline
- CHINA modern place name
- ● Upper Jurassic dinosaur site
- ○ Middle Jurassic dinosaur site
- ◉ Late Jurassic dinosaur site

Jurassic dinosaurs have been found in various parts of Gondwanaland (above), represented by good skeletons in parts of Africa, South America, Australia, India, and Antarctica, and by footprints from many other localities. These specimens all show similar faunas of megalosaurids and allosaurids (carnivores) and cetiosaurids and brachiosaurids (herbivores) which are like dinosaurs in other parts of Gondwanaland, and also to those of the northern hemisphere. Migration routes were open worldwide.

Piatnitkysaurus is similar to *Allosaurus* from North America, indicating that it was still possible for dinosaurs to cross between North and South America, even though the embryo North Atlantic Ocean had divided the two continents. The sauropods *Patagosaurus* and *Volkheimeria* are primitive types, similar to *Cetiosaurus* from the Middle Jurassic of England (see pages 92–93), further indicating faunal connections, perhaps through Africa, to Europe.

Isolated dinosaur bones were reported from Australia in the late 19th and early 20th centuries, and these include a partial sauropod skeleton, *Rhoetosaurus*, from the Early Jurassic. More recent finds from Australia have been mainly of Cretaceous dinosaurs (see pages 120–121). In the 1990s, a remarkable megalosaurid, *Cryolophosaurus* was found in Antarctica.

The main evidence that these Gondwanaland dinosaurs have given is that all continents were still accessible by a global and uniform dinosaur fauna during the Jurassic. The opening North Atlantic did not prevent mixing of faunas north and south. Equally, the straining among the various parts of Gondwanaland had not yet produced a major separation. Dinosaurs could walk readily from South America to Africa (the South Atlantic had not yet begun to open), but also across Antarctica to Australia, without hindrance. Collections in the Lower Cretaceous of Africa (see pages 122–123) confirm these links.

The Biggest Expedition of All

The German dinosaur-hunting expeditions to Tendaguru in Tanzania were the most spectacular of all.

The Late Jurassic dinosaurs of Tanzania are justly famous. The fauna is large, consisting of three meat-eaters and six plant-eaters, and they created a sensation when their skeletons were first exhibited in Berlin about 1920. The first finds were made in 1907 in what was then German East Africa at a locality called Tendaguru, four days march from the port of Lindi.

The locality was discovered by W. B. Sattler, an engineer working for the Lindi Prospecting Company, as he searched for mineral resources. He found pieces of gigantic fossil bones weathering out on the surface of the baking scrubland. Sattler reported back to the director of the company, and he in turn told Professor Eberhard Fraas, who happened to be in the colony at the time. A noted palaeontologist, A Fraas, had created his reputation describing Triassic fossil reptiles from Germany, and he had also worked on the early prosauropod *Plateosaurus* (see pages 76–77). Fraas visited the site, and recognised its huge potential. He collected some good specimens and in Stuttgart he showed them to various museum curators. Dr W. Branca, director of the Berlin Museum was enthusiastic, and he began raising funds for an expedition.

The Tendaguru site is located in the dry rough scrubland of Tanzania. Dr Werner Janensch (below) of the Humboldt Museum in Berlin led the Tendaguru expeditions in 1909, 1910, and 1911, during which his teams of workmen excavated spectacular numbers of bones of at least nine species of dinosaurs. He spent the rest of his life working on these specimens.

Branca, and his colleagues, succeeded in raising the huge sum of 200,000 marks enabling him to mount a substantial expedition, planned to run from 1909 to 1911, under the leadership of Dr Werner Janensch, curator of fossil reptiles at the Berlin Museum, and

The Tendaguru site is in Tanzania, East Africa (above), located more than 100 km inland from the seaport of Lindi (below right). Tendaguru is a small village, and a nearby hill, and the dinosaurs were found scattered over a wide area between the two.

The dinosaurs from Tendaguru include nine genera (below). The theropods include the small agile Elaphrosaurus and the larger Allosaurus and Ceratosaurus. The six genera of herbivores include the tiny ornithopod Dryosaurus, the stegosaur Kentrosaurus, and the sauropods Dicraeosaurus, Barosaurus, Tornieria, and the giant of them all, Brachiosaurus (see page 82). In addition, the Tendaguru fauna includes pterosaurs, fishes, and a tiny mammal, represented only by a jaw bone. The dinosaurs, and the rest of the fauna, are very similar to the animals from the Morrison Formation of the United States (see pages 100–103), indicating the ease of migration between North America and Africa at that time.

with the assistance of Dr Edwin Hennig. The expedition also ran in 1912, but under the leadership then of Dr Hans Reck.

The expedition was on a larger scale than anything mounted before. In the first season at Tendaguru, 170 native labourers were employed, and this rose to 400 in the second season, and 500 in the third and fourth. These huge numbers of workers were accompanied by their families, so the German dinosaur expedition at at Tendaguru involved an encampment of 700–900 people, creating enormous logistical problems.

The labourers dug numerous pits all over the site which ran from Tendaguru hill to Tendaguru village, a site spanning about 3 km. Most of the bones were very large, but they were fractured and had to be protected. The finds were mapped, measured, and then encased in plaster for then long journey to the coast. Teams of workmen carried the bones on their heads, or slung on poles, for the four-day trek to Lindi, where they were shipped out to Germany. In the first three years of the expedition, 4300 loads of fossil bones were sent out, weighing a total of 200 tonnes. During the fourth year, a further 50 tonnes were shipped out.

In Berlin, the long process of cleaning the bones began, and this lasted for many decades. As the materials were cleaned up, the huge skeletons were mounted in the Berlin Museum, where they may still be seen, and the new species were described up to the 1960s by Janensch, Hennig, and others. The Berlin specimen of *Brachiosaurus* the largest complete dinosaur skeleton in the world.

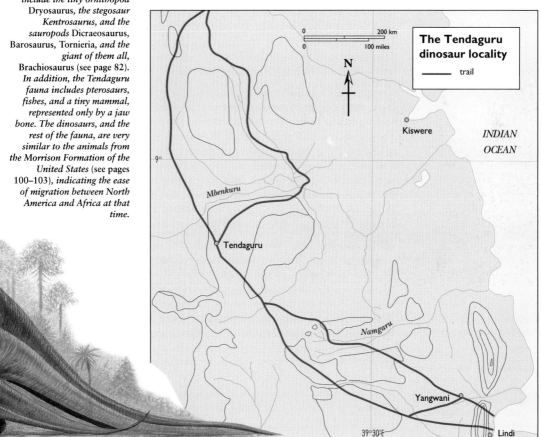

The Tendaguru dinosaur locality

—— trail

Dinosaurs of the Morrison Formation

The early dinosaur hunters in the wild west faced great hardships in their race to uncover the great American dinosaurs.

Bone hunters of the wild west (above). Here, a young Henry Fairfield Osborn, with the rifle, poses with Francis Speir (centre) and William Berryman Scott (right) in 1877, when Osborn was 20. These men were working in the Bridger Basin, searching for Tertiary mammals. Osborn later extended his interests to Morrison dinosaurs, and he sent expeditions to Colorado in 1897 and 1898 in search of bones of giant sauropods.

The first major dinosaur locality in the American West came to light by chance in 1877, but it provided a spur to the efforts of two highly active and ambitious palaeontologists who were based on the east coast, Othniel Charles Marsh (1831–1899) and Edward Drinker Cope (1840–1897) (see pages 24–25). By 1877, Cope and Marsh were highly antagonistic to each other after a number of disputes over east-coast fossils. In that year, two schoolmasters found some gigantic bones in the Morrison Formation of Colorado. One of them, Arthur Lakes, contacted both Marsh and Cope. He sent some bones from his site at Morrison to each man, and described the locality and how rich in bones he thought it might be. Neither Cope nor Marsh knew that the other had been contacted.

Marsh sent Lakes $100, and asked him to keep the discovery a secret, but it was too late for that. Cope had already begun describing the bones that had been sent to him for publication in the *Transactions of the American Philosophical Society*. Then Lakes, as a result of Marsh's cash and his request for secrecy, asked Cope to send the bones to Marsh. This must have been an infuriating request for Cope, but Marsh had already beaten him, and he had made arrangements with Lakes for excavation at Morrison. Marsh's victory was short-lived however, since the other schoolmaster, Mr O. W. Lucas, had sent further dinosaur bones to Cope, collected at Garden Park, near Canyon City in Colorado, also in the Morrison Formation. To Cope's delight the specimens from his locality at Garden Park were bigger than the Como Bluff specimens, and they were better preserved.

Cope's victory was also of brief duration, since Marsh then received news from two men in Wyoming, Messrs 'Harlow and Edwards', that they had some gigantic bones for sale. Marsh asked them to send their bones to New Haven, and he sent them $75 for their troubles, but he was surprised to be told that they could not cash his cheque. Marsh sent his assistant, Samuel Wendell Williston to investigate the discovery, and the mysterious 'Harlow and Edwards'. He found that these were pseudonyms for Mr W. E. Carlin and Mr Bill Reed, both employees of the Union Pacific Railroad, and Marsh approved of this secrecy, since he knew how to keep his discoveries confidential until the last possible moment in order to confound his rival Cope.

Carlin and Reed had found their fossils at Como Bluff, a long east-west-running ridge in southern Wyoming close to the new railroad they were building. Williston informed Marsh that the bones "extend for *seven* miles and are by the ton...The bones are very thick, well preserved, and easy to get out". The last observation pleased Marsh, as did Williston's later observation that "Canon City and Morrison are simply nowhere in comparison with this locality both as regards perfection, accessibility and quantity". Williston waited in an agony of suspense in case news of the discovery should leak out, and very speedily Marsh drew up an agreement for Carlin and Reed to sign, by which they agreed to work exclusively for him and to exclude all other collectors from Como Bluff, for a fee of $90 each per month. Carlin and Reed were active men, and they excavated bones at a great rate from 1877 onwards.

Palaeogeography of the Morrison Formation (far right). Sediments fed down from the proto-Rockies on to broad plains on the shores of great lakes and seas. These were the stamping grounds of the great Morrison dinosaurs.

Cope, of course, soon heard of the amazing new site at Como Bluff, and he sent out field parties to the area in 1879 and 1880. There are numerous leg-

Palaeogeography of the American Midwest in the Late Jurassic

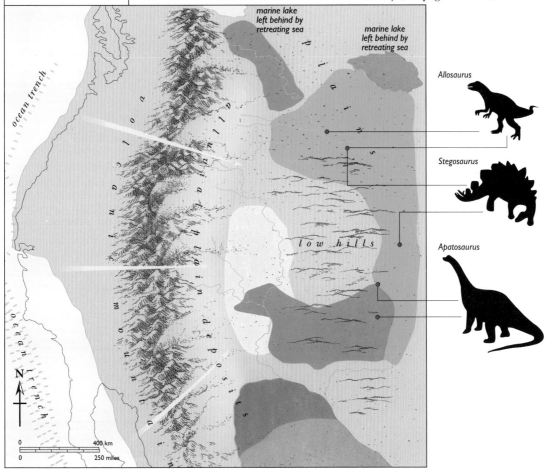

- ancient continent
- prevailing winds bringing ash deposits
- modern coastline
- • major dinosaur find

Late Jurassic deposits:
- river sandstones
- river silts and shales
- coarse conglomerates
- marine limestones
- salt beds (salt lake deposits)
- marine sandstones

ends of pitched fights between the Marsh and Cope workmen, but no evidence for such events. However, Reed was constantly on the lookout for Cope's workers, and reported their activities to Marsh in detail. In return, Marsh sent his men to Garden Park to report on Cope's excavations, and to make additional collections there. The Como Bluff dig continued until 1889, run by different Marsh employees after Reed left in 1883.

Marsh and Cope, at the same time as their massive operations in the Morrison Formation, also funded major dinosaur digs in the Late Cretaceous of Montana and Colorado, as well as excavations of mammals in the Tertiary beds of the same parts of the American West. Cope died in 1897, and Marsh two years later, and so ended an amazing period of rivalry.

Work continued on the Morrison Formation, however. A new generation of palaeontologists moved in, and they continued to make superb discoveries, although the old rivalries had gone, and the rough and ready excavation teams were replaced by more professional outfits. Henry Fairfield Osborn (1857-1935), later to become the powerful director of the American Museum of Natural History, went as a young man to work near Como Bluff in 1897. In 1898, his team moved north of Como Bluff, and they stumbled upon a hillside that was covered with dinosaur bones. So abundant were the bones that a shepherd had built himself a cabin from them; the site became known as Bone Cabin Quarry. Further dramatic discoveries were made in the Morrison Formation in the twentieth century (see pages 104-105).

marine lake left behind by retreating sea

marine lake left behind by retreating sea

Allosaurus

Stegosaurus

Apatosaurus

ocean trench

ocean trench

low hills

N

0   400 km

0   250 miles

Dinosaurs of the American Midwest

The Morrison Formation, covering much of the American Midwest, has proved to be the richest single source of dinosaur bones.

"His shotgun was presently heard and after a somewhat toilsome walk in the direction of the sound I heard him shout. I came up to him standing beside the weathered-out femur of a Diplodocus lying at the bottom of a very narrow ravine into which it was difficult to descend... there it was, as clean and perfect as if it had been worked out from the matrix in a laboratory."

W. J. Holland, describing Earl Douglass' find of dinosaurs in 1908 at what is now Dinosaur National Monument.

Stegosaurus (right) is one of the best-known dinosaurs from the Morrison Formation. This herbivore is famous for its tiny brain: in a dinosaur measuring 6-7m in length, it had a brain no larger than a kitten's. The bony plates may have been used for defence or for temperature control.

The wealth of dinosaurian fossils in the Late Jurassic Morrison Formation was revealed by the astonishing efforts of dinosaur hunters funded by Marsh and Cope from 1877 to the 1890s (see pages 100–101). These early excavations had been motivated by a desire to excavate as many giant bones as possible in the shortest possible time, and the men who did the digging, and their methods, were often extraordinarily crude. New approaches were introduced in 1897, when the first American Museum expedition entered the area, and work on the Morrison Formation this century has revealed a great deal about the life of the Late Jurassic.

The American Museum of Natural History continued operations in the Morrison Formation until 1905, and during that time sent many tonnes of bones back to the fledgling museum in New York. These provided a basis for one of the world's best dinosaur collections, and many of the Morrison dinosaurs collected around 1900 form the core of present exhibitions.

Other museums were looking towards the Morrison Formation about 1900 as a sure source of spectacular dinosaur specimens to form the core of their displays. The Carnegie Museum had been set up in Pittsburg from donations by Andrew Carnegie, a Scotsman who had made his fortune making steel. Earl Douglass, on the staff of the Carnegie Museum, first devoted his efforts to collecting Tertiary mammals in Montana and Utah. In 1908, he was visited in the field by W. J. Holland, the director of the Carnegie Museum, and he suggested to Douglass that they should perhaps look at some Jurassic rocks in the vicinity. This they did, and very soon Douglass stumbled upon a perfect *Diplodocus* femur, lying isolated at the bottom of a ravine.

The following year Douglass returned to the locality, in arid canyonlands close to the western border of Colorado with Utah. His brief was to find the precise source of the *Diplodocus* femur, and to assess whether the site could be excavated. After many days of prospecting, marching up and down the dry canyons, Douglass hit paydirt. He spotted some vertebrae of a big dinosaur in articulation, and this proved that this was an undisturbed dinosaur bonebed. Douglass wrote to his director, Holland, and together they excavated the skeleton represented by the vertebrae, later to be named *Apatosaurus louisae* in honour of Carnegie's wife.

With this encouragement, Carnegie donated the dollars necessary for large-scale excavations. Earl Douglass decided to remain at the site, and to work there full-time. He used the money to hire workmen locally and to buy heavy equipment for the taks of removing rock, and of moving the bones out. His wife, delightfully named Pearl Douglass, joined him in the remote corner of the Colorado-Utah border where the first bones were found, and the family remained there for many years. The nearest town, Vernal, Utah, was 30km away, so Douglass set up a homestead and built a log cabin right on the bone quarry. His excavations lasted from 1909 to 1923.

The meat-eater Allosaurus *is a typical Morrison dinosaur, known from thousands of bones collected at various localities in Colorado and Utah.*

During this time, Douglass and his crews excavated hundreds of skeletons, which are now in the Carnegie Museum, and even after so much effort, the bonebed showed no sign of running out.

Douglass unearthed skeletons of the sauropod *Apatosaurus*, the first find from the site, as well as the longer sauropod *Diplodocus*, the plated *Stegosaurus*, the bipedal ornithopod *Camptosaurus*, and the predatory *Allosaurus*. These dinosaurs had been found by Cope and Marsh before, but no site in the Morrison Formation had produced multiple skeletons of each form in a single locality. Douglass collected twenty complete skeletons, and isolated remains of a further 300 individual dinosaurs.

Douglass was at first puzzled by the immense richness of the site. How had such a diversity of dinosaur skeletons ended up in one place? He realised that these animals could not all have come here and died by some catastrophe: there were too many of them, and the range of species represented was too great. Douglass suggested that he had chanced upon an ancient sandbar that lay in the middle of some great meandering river in the Late Jurassic, and that the skeletons were accumulated by normal river action. Carcasses of animals that had died upstream were washed along, and eventually ran aground on the bar, where the flesh rotted, and where they were eventually buried under more sand brought down by the river.

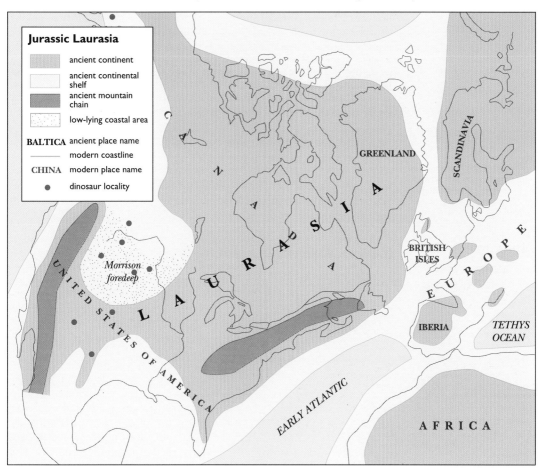

Jurassic Laurasia

- ancient continent
- ancient continental shelf
- ancient mountain chain
- low-lying coastal area

BALTICA ancient place name
— modern coastline
CHINA modern place name
● dinosaur locality

CANADA

GREENLAND

SCANDINAVIA

LAURASIA

UNITED STATES OF AMERICA

Morrison foredeep

BRITISH ISLES

EUROPE

IBERIA

TETHYS OCEAN

EARLY ATLANTIC

AFRICA

Megatrackways and Dinosaur Freeways

Footprints show how the dinosaurs migrated vast distances in search of food, and suggest complex herd structures.

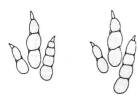

Dinosaur footprints typically have three toes. The other two toes were either held high off the ground or were lost during evolution and are therefore absent in later forms. Prints may be pressed deep into the sediment, often deforming several layers and forming tracks at the top and under-tracks beneath.

In recent years, palaeontologists have devoted serious efforts to the understanding of dinosaur footprints. The first of these were found in 1802, when Pliny Moody unearthed the specimen of a great three-toed print from the bedrock of his New England farm. He and his contemporaries thought that such specimens were the footprints of giant ancient birds that had perished in the Great Flood. Since then, thousands of prints have been reported in every part of the world, but they have often been seen simply as curiosities. New work has shown their true value.

The most exciting discoveries have been dinosaur megatrack sites, where footprint-bearing beds may extend for hundreds or even thousands of square kilometres. Palaeontologists have found such localities in Jurassic and Cretaceous rocks in North America in particular, but also on other continents. The sheer abundance of tracks can be staggering. One of the oldest sites, the Entrada Sandstone of eastern Utah, dated as Mid Jurassic, covers an area of about 300 square kilometres. Detailed mapping of small sections of this huge area show that the density of individual footprints averages between one and 10 per square metre. This gives a total of between 300 million and 3 billion dinosaur footprints at this one level, which may have been exposed at the land surface for tens or hundreds of years and thus represents a major hiatus in sedimentation. Over time, small populations of dinosaurs could make many millions of footprints.

Later, in the Late Jurassic and Early Cretaceous, track sites become abundant in North America. During much of the Cretaceous, North America was divided by the wide Western Interior Seaway, a sea which dinosaurs could not cross. Track sites range north and south along the western shore of this seaway, from Alaska to New Mexico, and the animals were always travelling parallel to the shore. In some places the tracks are abundant, showing that many animals passed, and some of these sites have been dubbed 'dinosaur freeways'. It is possible that these freeways give direct evidence of long-distance migrations. The dinosaurs may have headed north in the summer to feed on the lush vegetation, and south in the winter to avoid darkness and cold.

Below: A detailed map of a multiple dinosaur trackway dating from the Lower Cretaceous of New Mexico. Numerous small ornithopod dinosaurs (prints shown in red) have moved mainly from the bottom to the top of the rock as shown. A larger ornithopod (prints shown in black) has come in from the top and moved downwards.

Tracks can also provide detailed information about the behaviour of dinosaurs. The shape of a footprint shows which kind of dinosaur was responsible and how large it was, and the spacing of the prints shows how fast it was moving (prints which are close together indicate slow speeds, while well-spaced prints indicate that the animal was running). Multiple tracks show that a herd had passed by, and some of these show that small animals travelled in the middle of the herd while adults travelled on the outside, presumably in order to offer protection. Other tracks may show dinosaurs entering a lake and swimming or perhaps paddling on the bottom with their hind legs only.

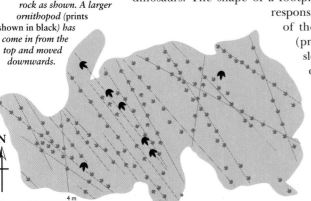

N

0   4 m

0   4 yds

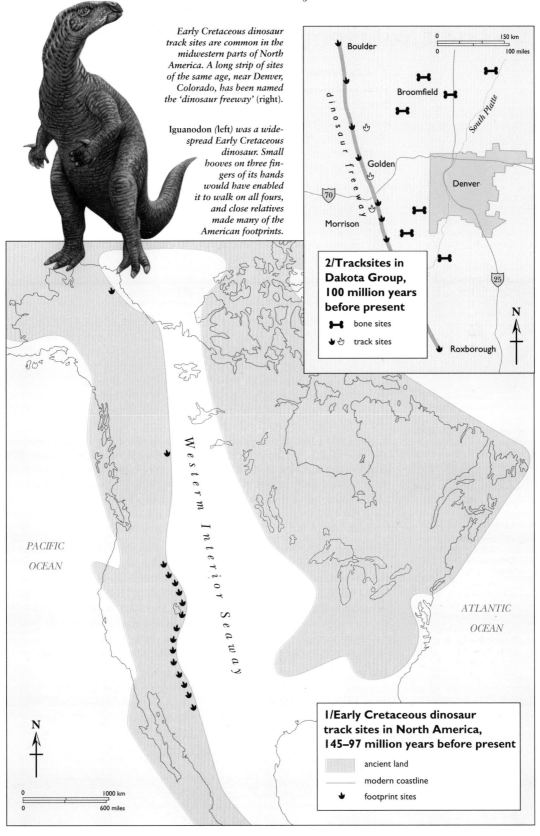

Early Cretaceous dinosaur track sites are common in the midwestern parts of North America. A long strip of sites of the same age, near Denver, Colorado, has been named the 'dinosaur freeway' (right).

Iguanodon (left) was a widespread Early Cretaceous dinosaur. Small hooves on three fingers of its hands would have enabled it to walk on all fours, and close relatives made many of the American footprints.

Boulder

Broomfield

South Platte

dinosaur freeway

70

Golden

Denver

Morrison

25

2/Tracksites in Dakota Group, 100 million years before present

- bone sites
- track sites

N

Roxborough

0   150 km
0   100 miles

PACIFIC OCEAN

Western Interior Seaway

ATLANTIC OCEAN

N

0   1000 km
0   600 miles

1/Early Cretaceous dinosaur track sites in North America, 145–97 million years before present

- ancient land
- modern coastline
- footprint sites

V: The Cretaceous: new dinosaur groups

The Cretaceous, a period lasting some 80 million years, witnessed fresh evolution among the dinosaurs. The Jurassic sauropods, stegosaurs, and theropods largely disappeared, and the world was populated by new groups, ornithopods, pachycephalosaurs, ceratopsians, ankylosaurs, and extraordinary new flesh-eaters such as the intelligent fast-moving ornithomimids and troodontids and the giant tyrannosaurids.

The Cretaceous world was warm and humid, as the Jurassic had been, but major changes took place among plants and insects. The changes in the plants may have had critical effects on dinosaurian evolution. In the seas too, a revolution was underway, with new kinds of seabed dwellers, new kinds of fishes, and new marine predators. All these changes were setting the scene for the modern world, a world still populated highly successfully by the dinosaurs.

In Cretaceous Seas: the Mesozoic Marine Revolution

The animals of seabed and open water that had characterised the Jurassic (see pages 78–83) continued into the Cretaceous. But there were new predators which had new methods of hunting. The whelks and other gastropods, molluscs that generally have coiled shells, had perfected their hole-boring techniques. Gastropods have a rasping tongue-like structure called a radula, covered in small horny teeth, which they use to scrape up plant material and other food from the rocks. Slugs and snails use the radula to tear and snip at cabbages and other garden plants. In the Cretaceous, several groups of gastropods began to exploit a new kind of carnivorous diet. They used their radulae to bore holes in a variety of shelled animals, usually bivalves, such as clams, oysters, and mussels, but also sea urchins, corals, and other sea-bed animals with calcareous shells. Added to the rasping effect of the radula, many of them used dilute acids. The acids eat away the calcite of the shell of the prey, and a hole appears. The gastropod then inserts a mouth tube, and slurps out the living body of its unfortunate victim.

Gastropods, like this whelk, are voracious predators. The siphon, right, is a feeding and breathing tube. The toothed tongue, or radula, is used for feeding on a variety of plants and animal prey.

The history of this delightful mode of feeding by shell-boring can be traced in the fossil record with precision: the boreholes are preserved in the shells, and it is even possible to identify the predator by the exact shape of the borehole. Over time, it can be shown that gastropods got better at the technique, by counting up the number of unsuccessful holes. An unsuccessful borehole is one that does not go right through the shell, and indicates that the gastropod must have given up before it reached the bivalve flesh. The ratio of successful to unsuccessful boreholes improves through geological time

As if that was not bad enough for the humble bivalve, new arthropod groups appeared, distant relatives of the Palaeozoic trilobites (see pages 39–41), the crabs and lobsters. These familiar animals have massive claws which they use to snip and break up hard shells. Cretaceous fossils often include shells which have been neatly clipped open by a crab, using an effect something like an old-fashioned can opener. Again, the bivalve has nowhere to hide, as the crab strips its protective shell away in surgical fashion. Another mode of attack came from starfish, which wrapped themselves over shells, and, using numerous suckers under their arms, gripped the shells, and slowly prised them open. The starfish then effectively turns its stomach inside out into the open bivalve shell, and digests the animal in its own home. Finally, there were new groups of fishes that exploited new predatory techniques to catch swimming prey, and to attack sedentary, but armoured, prey.

A major radiation of bony fishes occurred in the Late Jurassic and Cretaceous, following on from earlier radiations in the late Palaeozoic and early Mesozoic (see pages 80–81). The new bony fishes were the teleosts, the most diverse and abundant fishes today, including 20,000 living species, such as eel, herring, salmon, carp, cod, anglerfish, flying fish, flatfish, seahorse, and tuna. The huge success of the teleost radiation may be the result of the remarkable teleost jaw apparatus. Earlier bony fishes opened their jaws in a simple manner, as we do, but teleosts can project the whole jaw apparatus like an extendable tube. This came about because of a great loosening of the elements of the skull: as the lower jaw drops, the tooth-bearing bones of the upper jaw (the maxilla and premaxilla) move up and forwards. Rapid projection of a tube-like mouth allows many teleosts to suck in their prey, while others use the system to vacuum up food particles from the sea floor, or to snip precisely at flesh or coral. These fiendish new feeding techniques

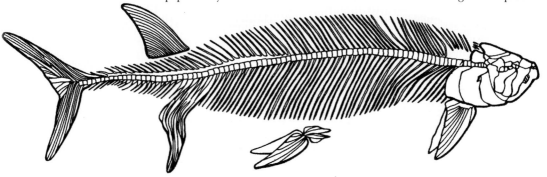

Xiphactinus, one of the largest teleosts, was 4–5m long. It preyed on other fishes in the Late Cretaceous seas of North America.

were a further threat to seabed life in the Cretaceous.

Some early teleosts became very large. *Xiphactinus*, for example, reached a length of 4 or 5 metres. One specimen has been found in the Late Cretaceous of North America with a 1.6–metre long fish in its stomach area, and others have been preserved with up to ten recognisable fish skeletons inside. These were clearly voracious predators! Teleosts built on their huge success after the Cretaceous, and they diversified enormously from that time to the present day.

The rise of such new and deadly predators in the Cretaceous, the hole-boring gastropods, the shell-snipping crabs and lobsters, the shell-gripping starfish, and the fast-swimming teleosts, has often been termed the Mesozoic Marine Revolution. The prey animals obviously did not all succumb to the onslaught of new predators, although some did. The bivalves adopted two defensive strategies. The most successful was to burrow deeply into sediment. Many bivalves in shallow water form burrows many centimetres deep, and only their siphons, tubes for pumping water and food particles through the shell, appear on the surface of the sand. Other bivalves, like the scallops, became swimmers themselves. Scallops sit quietly on the sea bed, but when threatened they clap their shells together, and shoot backwards rapidly out of reach of the predator.

Other marine animals defended themselves in different ways. Jellyfish evolved stinging cells, which can sometimes be fatal to fishes (and humans), sea urchins evolved long spines, partly for moving around, and partly to make their body portion harder to get at, and some sea urchins also became burrowers. The belemnites and ammonites, abundant shellfish of Jurassic and Cretaceous seas (see pages 78–79), could escape many of the new preda-tors since they were already capable of fast escape-swimming by rapid squirts from their siphons. Some of them were probably also able to make ink, as octopus and squid do today, and they could create confusion when attacked by squirting out a dusky cloud.

Predatory Marine Reptiles and Birds

Several groups of four-limbed vertebrates returned to the water during the Mesozoic, and the rise of the teleosts, as well as the abundance of ammonites and other shellfish, seems to have tempted a whole range of reptiles and birds to seek their food in the sea. Larger marine predators included ichthyosaurs, plesiosaurs, and mosasaurs. Cretaceous ichthyosaurs were simi-lar to those of the Jurassic (see pages 81–82), but the group dwindled in importance, and the ichthyosaurs died out in the Late Cretaceous, 30 Myr before the end of the dinosaurs and other great reptiles. The plesiosaurs, also common in the Jurassic (see pages 82–83), continued to evolve along two lines. The long-necked plesiosauroids included some extraordinary Late Cretaceous forms, the elasmosaurs, which had enormously long necks, some with over 70 neck vertebrae. The pliosaurs were also important top predators of Cretaceous seas, feeding on large fishes, as well as on ichthyosaurs and smaller plesiosaurs. There were four further marine predatory groups in the Late Cretaceous: turtles, giant lizards, pterosaurs, and birds.

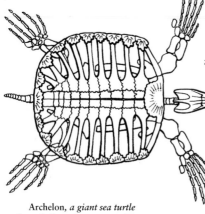

The turtles had arisen in the Late Triassic, and the group had a modest history through the Jurassic, remaining mainly terrestri-al. However, in the Late Cretaceous, several turtle groups entered the sea as fish-eaters. The shoals of teleost fishes may have offered a new food source. Some of the Late Cretaceous marine turtles, such as *Archelon* from the mid-American seaway of North America, reached a length of 4 metres, rather larger than typ-ical marine turtles today. The leatherback is up to 2 metres long and weighs 500 kg, so *Archelon* probably weighed over 1 tonne.

The mosasaurs, long serpentine animals with massive pointed jaws, fed on ammonites and fishes. Although they were often very large, and they all lived in the sea, the mosasaurs were lizards, giant relatives of the living monitor lizards. Mosasaurs are known only from the Late Cretaceous, but they became immensely abundant in the chalk seas of Europe and North America (see pages 116–117), with representatives also in Africa.

Archelon, a giant sea turtle from the Late Cretaceous mid-American seaway. This monster was up to 4m long, and it may have weighed 1 tonne.

The third group of marine predators in the Cretaceous were the pterosaurs. Pterosaurs had arisen in the Triassic, and they had radiated during the Jurassic (see pages 84–85). During the Cretaceous, pterosaurs diversified to exploit a wide range of diets, and they also became large. Typical pterosaurs had long spaced teeth for piercing and holding fish which they caught by trawling their lower jaws through the seawater. *Pterodaustro* from the Lower Cretaceous of South America has 400–500 slender flexible teeth in each jaw, which were probably used to filter microscopic plankton from the water. The teeth would have acted as a fine filter mesh in trapping thousands of small organisms which could be licked off and swallowed. The jaws of *Pteranodon* from the Late Cretaceous of North America are deep and hatchet-shaped and bear very few, or no, teeth. These forms also probably fished by beak trawling, and swallowed their catch so rapidly that no teeth were needed.

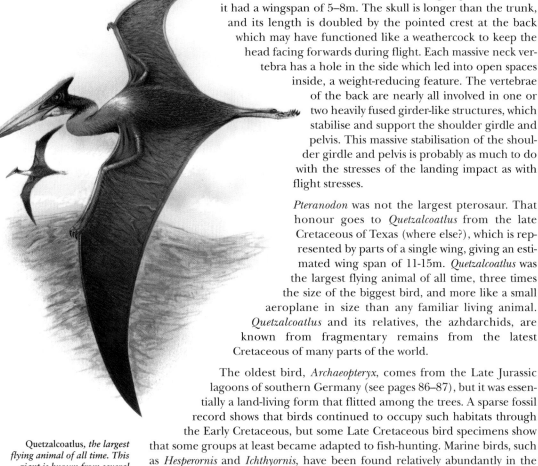

Pteranodon is one of the best-known and largest pterosaurs, and it had a wingspan of 5–8m. The skull is longer than the trunk, and its length is doubled by the pointed crest at the back which may have functioned like a weathercock to keep the head facing forwards during flight. Each massive neck vertebra has a hole in the side which led into open spaces inside, a weight-reducing feature. The vertebrae of the back are nearly all involved in one or two heavily fused girder-like structures, which stabilise and support the shoulder girdle and pelvis. This massive stabilisation of the shoulder girdle and pelvis is probably as much to do with the stresses of the landing impact as with flight stresses.

Pteranodon was not the largest pterosaur. That honour goes to *Quetzalcoatlus* from the late Cretaceous of Texas (where else?), which is represented by parts of a single wing, giving an estimated wing span of 11-15m. *Quetzalcoatlus* was the largest flying animal of all time, three times the size of the biggest bird, and more like a small aeroplane in size than any familiar living animal. *Quetzalcoatlus* and its relatives, the azhdarchids, are known from fragmentary remains from the latest Cretaceous of many parts of the world.

The oldest bird, *Archaeopteryx*, comes from the Late Jurassic lagoons of southern Germany (see pages 86–87), but it was essentially a land-living form that flitted among the trees. A sparse fossil record shows that birds continued to occupy such habitats through the Early Cretaceous, but some Late Cretaceous bird specimens show that some groups at least became adapted to fish-hunting. Marine birds, such as *Hesperornis* and *Ichthyornis*, have been found relatively abundantly in the Late Cretaceous limestones of the mid-American seaway in Kansas and neighbouring states. These birds still had teeth, but the bony tail was reduced to a short knob as in modern birds, and they had other modern features. *Hesperornis* was a large loon-like diving bird, while *Ichthyornis* was a flier that flapped over the waves, and submerged itself to chase fish. Both forms probably used webbed feet to power themselves through the water.

Quetzalcoatlus, *the largest flying animal of all time. This giant is known from several partial skeletons from Texas, and its wingspan is estimated to have been 11–15m, perhaps 12m. It is not certain what* Quetzalcoatlus *fed on whether fish caught at sea, or dinosaur carcasses, a source of scavenged food.*

Coevolution of Flowering Plants and Insects

There were dramatic changes in plants during the Cretaceous. At the beginning of that time, the primitive plant groups dominated, conifers, seed ferns, bennettitaleans, horsetails, and ferns, and these formed the diet of the herbivorous dinosaurs. However, a major change happened during the Early Cretaceous, when a new group, the angiosperms, or flowering plants, came on the scene. Angiosperms today include most food plants, from cabbages to apple trees, herbs to coconut palms, as well as familiar plants like the broadleaved trees and the grasses.

The fossil record shows that the angiosperms arose during the Early Cretaceous, and some of the oldest specimens of flowers were reported in 1996 from the Wealden of England (see pages 118–119). Older records of angiosperms date from the Triassic, nearly 100 Myr earlier, but these have yet

Flowering plants arose in the Early Cretaceous, and by late Cretaceous times, when dinosaurs were still abundant, modern plants flourished worldwide. The flowering plants, or angiosperms owed their success to the flower and its precise control of pollination.

to be confirmed. After their certain origin and early radiation, angiosperms rose rapidly to dominance in the Late Cretaceous. The last dinosaurs lived in a landscape that looked rather modern, since angiosperms like roses, magnolias, water lilies, laurel, and dogwood were to be found extensively (see pages 132–133).

The secret of the success of the angiosperms is the flower and the fully enclosed ovule. Special covering tissues protect the ovule from fungal infection, desiccation, and the unwelcome attentions of herbivorous insects. In angiosperms, pollen is borne on long stamens arranged around the centrally-placed ovary or ovaries. Pollen grains are transported, by insects or by the wind, to the stigma, a sticky surface above the ovule, and from it the pollen grains send pollen tubes to the ovules through which the sperm pass.

Insects had existed long before the Cretaceous, and some important groups radiated in the Carboniferous and Permian, including the spectacular large dragonflies of the coal forests (see page 45). A whole suite of new insect groups appeared in the Cretaceous, and their success has been linked to their coevolution, or joint evolution, with the flowering plants.

Flowering plants show many special adaptations for attracting insects: multicoloured petals, special fragrances, and supplies of nectar (sugar water). The evolution of angiosperm characteristics was paralleled by the evolution of major new groups of insects that fed from flowers and pollinated the flowers. Groups of beetles and flies that pollinate various plants were already present in the Jurassic and Early Cretaceous, but the hugely successful butterflies and moths, ants, bees, and wasps are known as fossils only from the Early Cretaceous onwards. All of these are closely linked to flowering plants generally as a source of food, but they all perform the function of spreading the pollen from flower to flower. Both sides benefit: the insect gets food, and the plant secures cross-pollination, essential to avoid inbreeding, where pollen self-pollinates the overies of the same plant.

The hymenopterans, bees and wasps, evolved substantially during the Cretaceous. The first hymenopterans to appear in the fossil record, the sawflies, had been present since the Triassic. Some fossil specimens have masses of pollen grains in their guts, a clear indication of their preferred diet. The sphecid wasps, which arose during the Early Cretaceous, had specialised hairs and leg joints that show they collected pollen. Other wasps, the vespoid wasps and the true bees, appear to have arisen in the Late Cretaceous.

The first angiosperms, such as magnolias, may not have had specialised relationships with particular insects, and may have been pollinated by several species. More selective plant-insect relationships probably grew up during the Late Cretaceous with the origin of vespoid wasps which today pollinate small radially symmetrical flowers. These kinds of specialised relationships are shown by increasing adaptation of flowers to their pollinator in terms of flower shape, and the food rewards offered, and of the pollinator to the flower. Evidence from the Late Cretaceous plant record shows that angiosperms such as roses then had specialised features that catered for pollinators that fed on nectar as well as pollen.

The coevolution of angiosperms and insects gave a major boost to insect evolution. Bees and wasps include many social species, those that live together in colonies with complex mutual interactions. The other main social insect groups, termites and ants, also arose during the Cretaceous, and their success may be the result more of predation on angiosperms, than co-evolution of pollinating behaviour. Nevertheless, the rise of those four insect groups

One of the most fearsome flesh-eating dinosaurs, Deinonychus (below), from the Early Cretaceous of North America. This was a dromaeosaurid, or 'raptor', a modest-sized dinosaur that may have hunted larger prey in packs. It attacked by slashing with its fiendish toe claw.

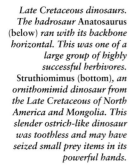

marked a major leap in total global diversity of insects and helped to drive the group to its present enormous diversity of many millions of species.

Cretaceous Flesh-eating Dinosaurs

The saurischian dinosaurs, including the meat-eating theropods and the long-necked plant-eating sauropodomorphs (see pages 72–73), had mixed fates in the Cretaceous. The theropods flourished, while the sauropodomorphs, immensely important in the Jurassic (see pages 83, 92–105), dwindled to unimportance in most parts of the world. The only significant sauropod family of the Cretaceous were the titanosaurids, known especially from the southern continents of Gondwanaland (see pages 124–125).

Among theropods, the ceratosaurs (see page 83) continued in the Cretaceous as the abelisaurids, another Gondwanan group (see pages 124–125). The tetanurans, on the other hand, blossomed and flourished. From ancestors like *Megalosaurus* and *Allosaurus* in the Jurassic (see pages 92–93, 102–103), tetanurans small and large evolved in the Cretaceous. Dromaeosaurids, such as *Deinonychus* from the Early Cretaceous of North America, were lightweight agile animals equipped with a massive claw on one of their toes on each foot. This claw was evidently used for slashing at prey animals, and it is possible that the dromaeosaurids hunted in packs, cutting and slashing at larger plant-eaters, until their intended prey fell to the ground, weakened by the attacks. The dromaeosaurids are the closest dinosaurian relatives of birds, sharing with them an elongate three-fingered hand, and numerous features of the rest of the skeleton (see page 86).

Late Cretaceous dinosaurs. The hadrosaur Anatosaurus (below) ran with its backbone horizontal. This was one of a large group of highly successful herbivores. Struthiomimus (bottom), an ornithomimid dinosaur from the Late Cretaceous of North America and Mongolia. This slender ostrich-like dinosaur was toothless and may have seized small prey items in its powerful hands.

The unusual toothless theropod *Oviraptor* from the Late Cretaceous of Mongolia (see pages 126–127) had earned itself a bad reputation, until recently. *Oviraptor* means 'egg thief', and that name was coined since the type skeleton was found in 1923 lying on top of a nest containing eggs. A further skeleton of *Oviraptor* was found in 1993, also located on top of a nest, but this time an embryo was found inside one of the eggs, and it turned out to be an unhatched *Oviraptor*. Far from being an egg thief, these *Oviraptor* individuals were apparently brooding their eggs.

The tyrannosaurids, such as *Tyrannosaurus*, the largest terrestrial carnivore

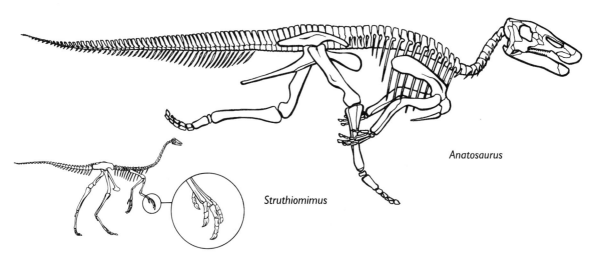

Anatosaurus

Struthiomimus

of all time at 14m long, radiated in North America and central Asia during the Late Cretaceous. *Tyrannosaurus* has a large head with an extra joint in the

The hadrosaurs were a diverse group of plant-eating dinosaurs in the Late Cretaceous. The different species of hadrosaur had an array of crests or no crests at all. These presumably acted as species recognition signals: each form had its own crest-shape, and this was a visual marker.

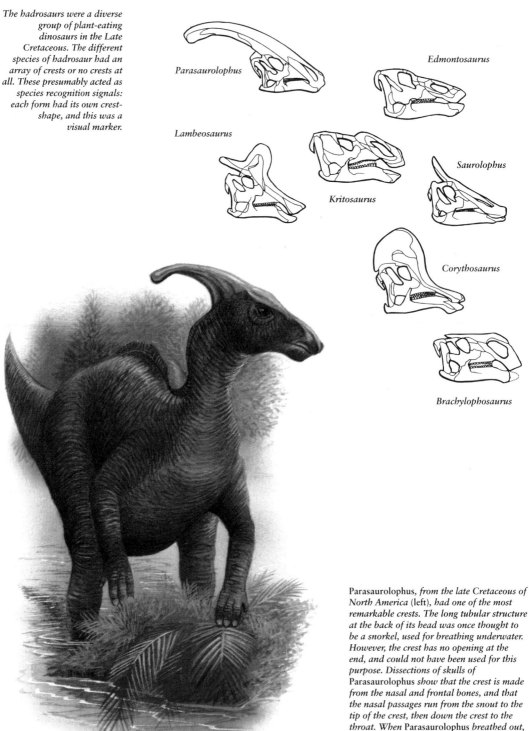

Parasaurolophus

Edmontosaurus

Lambeosaurus

Kritosaurus

Saurolophus

Corythosaurus

Brachylophosaurus

Parasaurolophus, from the late Cretaceous of North America (left), had one of the most remarkable crests. The long tubular structure at the back of its head was once thought to be a snorkel, used for breathing underwater. However, the crest has no opening at the end, and could not have been used for this purpose. Dissections of skulls of Parasaurolophus *show that the crest is made from the nasal and frontal bones, and that the nasal passages run from the snout to the tip of the crest, then down the crest to the throat. When* Parasaurolophus *breathed out, it honked. Each hadrosaur could make a different tooting or honking sound.*

lower jaw between the dentary and the elements at the back. This allowed *Tyrannosaurus* to increase its gape for biting large prey. Tyrannosaurids had tiny forelimbs equipped with either two or three fingers, but these would seem to have been be quite useless since they could not even reach the mouth. They may have been used to help *Tyrannosaurus* get up from a lying position, by providing a push while the head was thrown back and the legs straightened. Another recent idea is that they were used by males to stroke the females into a receptive mood during courtship.

The troodontids from the Late Cretaceous of North America and Mongolia also also belong in this group. *Saurornithoides* (see page 133) has a long slender skull with its eyes facing partly forwards so that it may have had binocular vision. The braincase is bulbous and relatively large, which has led to the interpretation of the troodontids as the most intelligent (or least stupid?) dinosaurs. Close relatives were the ornithomimids of the Late Jurassic to Late Cretaceous, highly specialised theropods with a slender ostrich-like body and long arms and legs. The hands have three powerful fingers which may have been used for grasping prey items. The lightly-built body indicates that *Struthiomimus* could have run fast, and speeds of 35–60km/h have been estimated. The skull is completely toothless in later forms, and the ornithomimid diet may have included small prey animals such as lizards or mammals, or even plants.

Ornithopods, Pachycephalosaurs, and Ceratopsians

The main plant-eating dinosaurs of the Cretaceous were ornithopod dinosaurs. The sauropods had dwindled in most areas, possibly as a result of the switch from the plants they favoured to the new angiosperms. The Cretaceous plant-eating dinosaurs, predominantly the ornithopods, as well as pachycephalosaurs and ceratopsians, all had a very different system of teeth that was apparently better adapted for dealing with large amounts of plant food efficiently. They had multiple rows of teeth inside the jaws, and when teeth at the top became worn, new ones shifted into position. A fourth important ornithischian group in the Cretaceous were the ankylosaurs, heavily armoured forms (see page 135) that are related to the stegosaurs (see pages 84, 94, 102).

Ornithopods arose in the Late Triassic, and early forms are known especially from the Early Jurassic of southern Africa (see pages 90–91). In the Early Cretaceous, the hypsilophodontids lived nearly worldwide, from the Wealden of England (see pages 118–119) to the south-pole regions of Australia (see pages 120–121). Iguanodontids were another highly successful group of the European Wealden, and the group includes *Ouranosaurus* from the Early Cretaceous of North Africa, which had spines on its back, perhaps supporting a sail for thermoregulation (see pages 122–123).

The most diverse, and most successful, ornithopod clade were the hadrosaurs or 'duck-billed' dinosaurs of the Late Cretaceous (see pages 126–133). Frequently, three or four distinct hadrosaurian species are found side by side in the same geological formation, and it seems evident that large mixed groups roamed over the lush lowlands rather as closely related antelope do today in Africa. The hadrosaurs are famous for their expanded duck-like bills and their massive dental batteries, which allowed them to process tough plant food. Hadrosaurs, like other ornithopods, had a chewing mechanism which may have contributed to their success. Reptiles do not chew their food, but merely swallow it and hope for the best. Mammals, such as ourselves, rely on chewing to ensure that the food is partially digested and ready for efficient extraction of nutrients when we swallow it. Hadrosaurs had a hinge above the cheek region of the skull, and when the jaws closed, this region moved outwards, and the plant food was ground with a strong sideways shearing movement.

Hadrosaurs all have essentially the same skeletons and skulls, but some have an impressive array of headgear. The premaxillae and nasal bones extend up and backwards to form in some a high flat-sided 'helmet', either low or high, square or semicircular, in others a long 'tube', spike, or forwards-directed rod. It has long been realised that the nasal cavities extended from the nostrils and into these crests, but it was once assumed that they acted as 'snorkels', especially in *Parasaurolophus*. However, this is impossible since there is no opening at the top of the crest. Air breathed in or out through the nose had to travel round this complex passage system.

The most likely function of the crests is as visual species and sexual signalling devices. Just as modern birds use colourful and often elaborate patterns of feathers to recognise potential mates, and to signal their position in dominance hierarchies, so too did the hadrosaurs, but with cranial crests. Seemingly, males and females of the same species had rather different crests. Further, David Weishampel has shown that the hadrosaurs augmented their visual display with an auditory one too. The shapes of the air passages within the crests, particularly those of forms like *Parasaurolophus*, are very like musical wind instruments. A powerful snort would create a low resonating note, and the shape of the air passages in males and females, and in juveniles, would give a different note. Species differences would have been even more marked. We can imagine the Late Cretaceous plains of Canada and Mongolia reverberating to deep growls and blaring squawks as the hadrosaurs went about their business!

The pachycephalosaurs, a small group of mainly Late Cretaceous herbivores from North America and central Asia, are characterised by their remarkably thick skull roofs. The parietal and frontal bones are fused into a great dome in some forms with the bone up to 22 cm thick. The pachycephalosaurs may have used their thickened heads in butting contests when seeking mates, as is seen today among wild sheep and goats. The pachycephalosaur, a biped, adopted a horizontal-backbone posture during the charge so that the force of the impact ran straight round the skull margins and down the neck to the shoulders and hindlimbs.

The Ceratopsia (literally 'horned faces') are a related larger group of about 20 genera known mainly from the Late Cretaceous of North America (see pages 130–133). All are characterised by a triangular-shaped skull when viewed from above, an additional beak-like rostral bone in the midline at the tip of the snout, a high snout, and an expanded frill over the neck region. *Protoceratops* from the Late Cretaceous of Mongolia and China (see pages 126–127) was a quadruped with the beginnings of a nose horn, a thickened bump in front of the orbit. Ceratopsian skulls show great variations in their frills and horns: some have a simple nose horn, while others have 'horns' below the eyes, while *Triceratops* has long horns above the eyes (see page 133). The frill may be short or long, and indeed *Torosaurus* had a 2.6m long skull in which the frill is longer than the skull itself, the largest skull known from any land animal. The frills and horns may have been used in defence and as visual species-signalling structures as well as in threat displays. Male ceratopsians may have engaged in head wrestling with the horns interlocked, just as deer do today

Confrontation in the Late Cretaceous (opposite). A herd of plant-eating Triceratops is menaced by the awesome predator Tyrannosaurus. Perhaps the ceratopsians used their bulky bodies and head armour to form an impregnable shield around the younger and weaker members of the herd.

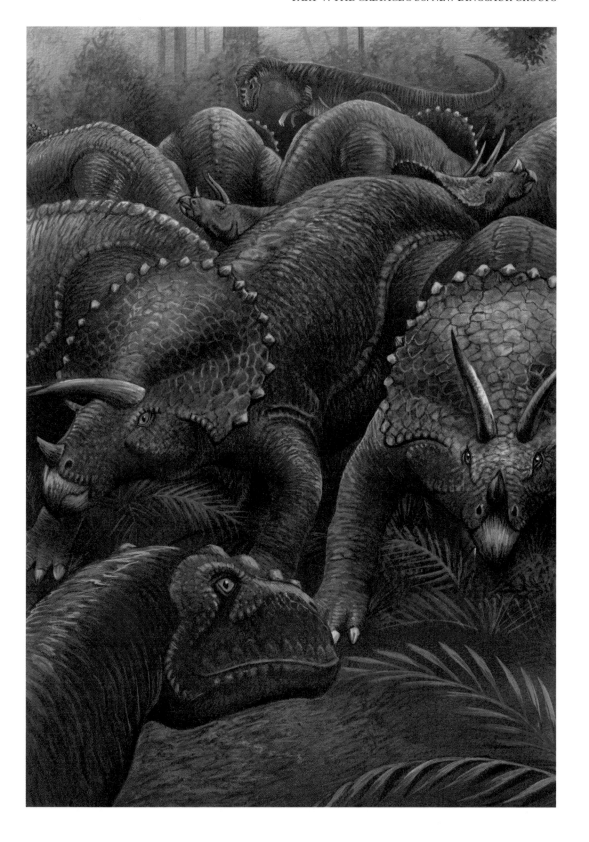

The Cretaceous World

Cretaceous climates continued warm and humid, and dinosaurs lived worldwide. The further break-up of Pangaea led to some endemism.

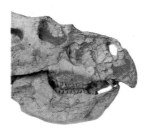

The skull of Protoceratops, *representative of one of the new Cretaceous dinosaur groups, the ceratopsians. Ceratopsians arose in the Mid Cretaceous, and they were important through the Late Cretaceous, right to the end of the period. These herbivorous dinosaurs had a bony extension, or frill, running from the back of the skull, over the neck, perhaps as a point of origin of enlarged jaw muscles, and perhaps as protection of the neck area.* Protoceratops *is known from the Late Cretaceous of Mongolia, where it was one of the commonest animals of its time.*

Cretaceous climates in many ways resembled those of the Jurassic. Dinosaurs of the Early Cretaceous are best known from the Wealden of Europe (see pages 118–119), and the Wealden sediments and plants indicate warm humid conditions. The same is true of the classic Late Cretaceous dinosaur localities of Mongolia (see pages 126–127) and North America (see pages 130-133). In all these cases, the richest finds of dinosaurs come from low-lying areas, close to the sea, where great river systems carried large volumes of sediments down from neighbouring mountains.

Sea levels were 25m higher than the present day in the Early Cretaceous. In the Late Cretaceous, during the Cenomanian stage, about 100 Myr ago, sea levels rose dramatically to 200m deeper than present levels. The enormous Cenomanian marine transgression had the effect of flooding huge areas of land, and of dividing continental land masses. In the Late Cretaceous, inland seas flooded up through midwestern parts of the United States, nearly reaching the Canadian border, and much of central and western Europe was under water.

In Europe, the great Cenomanian marine transgression produced a shift from the continental sediments of the Wealden to the Chalk. Chalk is made from the tiny limestone skeletons of diatoms, minute floating planktonic plants, deposited at the bottom of warm shallow seas. Huge thicknesses of white chalk accumulated on the northern shore of the Tethys Sea, in a broad strip running from Ireland and Britain to the Middle East, and this striking rock type gave its name to the Cretaceous period (*Creta* is Latin for chalk).

During the Mid Cretaceous, there was a dramatic increase in mid-oceanic ridge activity. The North Atlantic continued unzipping northwards from its early phase of rifting in the Late Triassic and Jurassic (see pages 88–89). In the Mid Cretaceous, the mid-Atlantic ridge extended right down into the South Atlantic, and a new ridge system became established across the Indian Ocean, separating Africa and India to the north from Antarctica and Australia to the south. Mantle magmas welled to the surface, creating new oceanic crust, and forcing the continental masses apart (see pages 50–51). The ridge systems rose high above the old ocean depths, and they lifted the neighbouring deep ocean floors with them. If you push up the floor of a bath that is full of water, the water will overflow, and that is just what happened in the Cenomanian. The deep ocean floors rose, and sea water rose worldwide by 200m, overflowing vast areas of the continents.

Cretaceous dinosaurs
Fossil dinosaurs from the Cretaceous period are found on all continents. The most fertile places for fossil finds are in the semi-arid regions of North America and Asia, where fossil bones are well preserved, and the rock is eroded sufficiently to expose them.

Ornithomimus
Medium-sized ornithischian, 3 to 4m (10ft), including a 2m (6ft) tail. The first remains were found in 1889 in Colorado.

Gallimimus
At 4m (13ft) long, this was probably the largest ornithomimosaur. Its hands may have been used for scraping at the soil. It was discovered in China.

Nodosaurus
This dinosaur was armoured with small bony knobs, and had powerful limbs to support its body weight. Skeleton finds are from western North America.

Acanthopholis
This 5.5m (18ft) long animal was protected by rows of oval bony plates and sharp spikes, which lined its back. Its remains were first found in Kent in 1864.

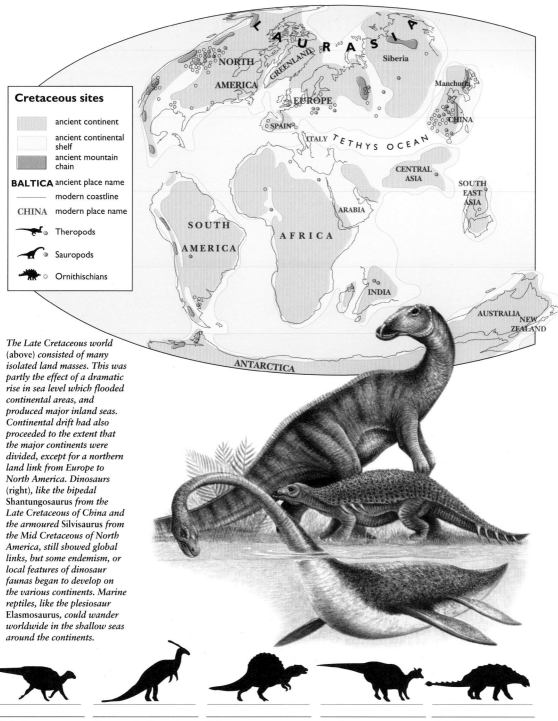

Cretaceous sites

	ancient continent
	ancient continental shelf
	ancient mountain chain
BALTICA	ancient place name
——	modern coastline
CHINA	modern place name
	Theropods
	Sauropods
	Ornithischians

The Late Cretaceous world (above) consisted of many isolated land masses. This was partly the effect of a dramatic rise in sea level which flooded continental areas, and produced major inland seas. Continental drift had also proceeded to the extent that the major continents were divided, except for a northern land link from Europe to North America. Dinosaurs (right), like the bipedal Shantungosaurus from the Late Cretaceous of China and the armoured Silvisaurus from the Mid Cretaceous of North America, still showed global links, but some endemism, or local features of dinosaur faunas began to develop on the various continents. Marine reptiles, like the plesiosaur Elasmosaurus, could wander worldwide in the shallow seas around the continents.

Brachylophosaurus
A duck-billed dinosaur; the first skull was found in Alberta, Canada in 1936. It was crested – possibly as a kind of identification signal for other dinosaurs.

Parasaurolophus
Remains of this duck-billed dinosaur have been found in western North America. A distinctive tubular crest, up to 1.8m (6ft) long, curved back from its snout.

Spinosaurus
This meat-eater was up to 12m (40ft) long. The spines on its back – possibly used to control body temperature – were 2 m high. It was found in North Africa.

Lambeosaurus
This crested duck-billed dinosaur reached a height of 15m (50ft). Crests in males were larger than in females. Remains have been found in western Canada.

Ankylosaurus
This dinosaur was over 10m (33ft) long, and protected by spines and bone plates. It had a club-shaped tail which could be used against attackers.

Dinosaurs of the Wealden

The Wealden of north-west Europe preserves the best Early Cretaceous dinosaurs. The discoveries include some of the first dinosaurs to be named.

The second and third dinosaurs to be named came from the Wealden of England. The first was *Megalosaurus* from the Middle Jurassic of Oxfordshire, named by Buckland in 1824 (see pages 22, 92–93). One year later, Gideon Mantell described the ornithopod *Iguanodon* from the Early Cretaceous of Sussex, and in 1832, he named the third dinosaur, *Hylaeosaurus*, an armoured ankylosaur, from the same locality.

The English Wealden provided rich pickings for the early collectors. The rocks covered a large area of the south-east of England, in the counties of Kent, Surrey, and Sussex, a region south of London known as the Weald, hence the name Wealden for the rocks. Wealden rocks with dinosaur remains were also found on the Isle of Wight. Bones came to light frequently in the Wealden of the Weald since there were many large sandstone quarries, and the quarrymen soon learned to look out for bones and keep them for the gentlemen naturalists, like Mantell, who would pay handsomely for specimens. Keen fossil-hunters soon turned to the Isle of Wight as well, where the seas of the English Channel were wearing away the cliffs on the west side of the island, and fresh dinosaur skeletons were (and are) exposed each year.

Reconstruction of Iguanodon *based on the famous Bernissart skeletons from Belgium. This reconstruction, produced by Louis Dollo in 1882, was one of the first based on a European dinosaur. It confirmed that many dinosaurs were bipeds, as had been seen from new finds in North America, and it solved a problem over the location of a conical pointed bone in* Iguanodon: *it was a thumb spike. This image of* Iguanodon *remained in use until the 1970s, when Peter Galton and David Norman realised that the ornithopod dinosaurs must have held their backbones more horizontally in order to achieve balance when they moved (see restoration opposite). Dollo had had to break the tail in order to make* Iguanodon *stand up like a kangaroo.*

The next highlight was the discovery in 1878 of a positive army of *Iguanodons* in a coal mine in Belgium. The bones came to light when miners were excavating a coal seam beneath the village of Bernissart. They saw remains of huge bones in the roof of one of their adits, and luckily found a few teeth as well. Palaeontologists from the Royal Museum of Natural History in Brussels moved in, and supervised the careful excavation of 39 skeletons of *Iguanodon*, most of them essentially complete, as well as the skeleton of a meat-eating megalosaurid, fishes, turtles, crocodilians, insects, and plants.

Louis Dollo (1857–1931) was given the job of sorting, preparing, and describing the astonishing Bernissart dinosaur collection. He supervised the cleaning of the skeletons, not an easy task, since the bones were damp, and cracked. The museum technicians had to use a terrifying cocktail of varnishes and glues to strengthen the bones, and this has led to endless conservation problems today. For the first time in Europe, Dollo was able to reconstruct some complete dinosaur skeletons in natural pose, efforts that rivalled those of Cope and Marsh in North America (see pages 24–25).

Dollo announced his results to the scientific world through a series of dozens of papers about the Bernissart *Iguanodons*, and the associated faunas. He was one of the first palaeontologists to consider the biology of the dinosaurs— how they lived—rather than merely giving them a name, and moving on rapidly to the next specimen, as was more the habit of his contemporaries. The new *Iguanodon* skeletons were more complete than anything yet found in England, and Dollo was able to solve a long-standing problem concerning the correct location of a conical pointed bone. Mantell thought this was a nose horn, and indeed Owen reconstructed *Iguanodon* in the 1850s with the bone mounted on the snout. The new skeletons from Bernissart showed that the mystery bone was a specialised thumb claw, used presumably for defence, or in fighting for mates.

A reconstructed Wealden scene (right) shows Iguanodon and a small herd of sauropods. By the Early Cretaceous, ornithopods had taken over as the main herbivores, and sauropods, which had ruled supreme until the Late Jurasic (see pages 98–103), declined in importance. Iguanodon was the commonest dinosaur of the Wealden, and it inhabited low-lying areas with broad meandering streams and lakes populated with fishes and crocodilians. Lush forests were scattered around, and Iguanodon could have fed on the coarse leaves of conifer trees, and on low waterside plants such as horsetails. Wealden rocks extended over southern England, Belgium, northern Germany, and parts of France (below).

Classic Wealden localities

●   major dinosaur finds

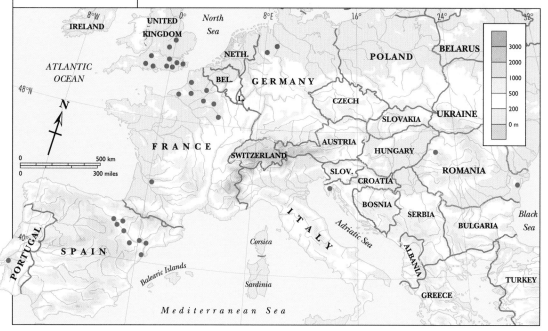

Polar Dinosaurs

New discoveries in Alaska and in Australia show that dinosaurs lived in polar regions. This has triggered a debate about their thermoregulation.

If dinosaurs were like modern reptiles, they should prefer to have lived in warm and dry environments. Modern lizards, snakes, turtles, and crocodilians favour warm climates since they are cold-blooded (ectothermic), and require external sources of heat to maintain their activity levels. Modern reptiles also are not so troubled by lack of water as are the warm-blooded (endothermic) birds and mammals since they are better adapted for water conservation. Endotherms need large volumes of water to operate their high metabolic rates and to assist in clearing waste products in urine. Reptiles produce solid urine (as do birds also), and hence water is not wasted.

The allosaurid from south Australia was a dramatic find, since the closest relatives of this form were known from the Late Jurassic of North America (see pages 102–103). The Australian allosaurid is represented by limited remains, but they seem diagnostic. This may be a late-survivor that lived on, isolated in Australia, long after the allosaurids had died out elsewhere.

The discovery of polar dinosaurs in the 1960s came as a shock. Oil geologists in northern Alaska, found dinosaur bones in Cretaceous sediments on the north slope. The north slope of Alaska also lay within the Arctic Circle in Cretaceous times. How could dinosaurs, if they were like modern reptiles, have lived in such polar conditions?

Arctic dinosaurs became part of the debate about dinosaur thermoregulation in the 1970s and 1980s. The proponents of dinosaurian endothermy argued that they must have been adapted for living in the cold. Only an endotherm can withstand long-term cold and Greg Paul, an enthusiast for endothermy in dinosaurs, pictured a hadrosaur picking its way through a snowdrift. Opponents of endothermy noted that there was only limited evi-

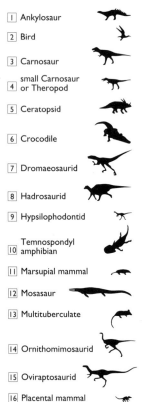

☐		
1	Ankylosaur	
2	Bird	
3	Carnosaur	
4	small Carnosaur or Theropod	
5	Ceratopsid	
6	Crocodile	
7	Dromaeosaurid	
8	Hadrosaurid	
9	Hypsilophodontid	
10	Temnospondyl amphibian	
11	Marsupial mammal	
12	Mosasaur	
13	Multituberculate	
14	Ornithomimosaurid	
15	Oviraptosaurid	
16	Placental mammal	

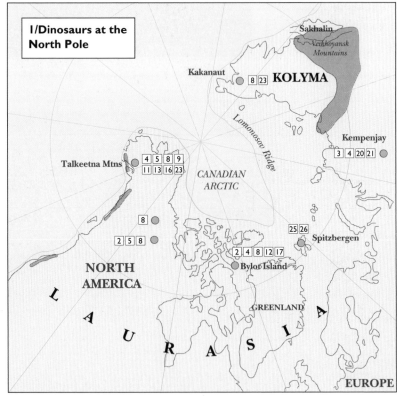

1/Dinosaurs at the North Pole

dence, if any, for ice and snow in Cretaceous Alaska, and that in any case, the dinosaurs could easily have migrated south during the cold dark winters. Like modern caribou in Canada, the dinosaurs would have followed the food supplies as the seasons changed, even if it meant trekking a couple of thousand kilometres each year.

More dinosaurs have been found in the Late Cretaceous on the north slope of Alaska since the 1960s. Specimens indicate a fauna of hadrosaurs and ceratopsians, as well as rarer flesh-eaters, just the same as at more southerly locations in Alberta and the midwestern United States. So far the Alaska sites lack many smaller amphibians and reptiles, such as frogs, lizards, and crocodilians. Exciting discoveries from the southern hemisphere in the 1990s have confirmed the existence of polar dinosaurs. A few dinosaur specimens have been found in the Jurassic and Cretaceous of Antarctica. The most dramatic south-pole dinosaurs come from finds by Pat and Tom Rich in the Early Cretaceous of Victoria, south Australia, a region that was located well within the Antarctic Circle in the Cretaceous. The fauna is not strikingly different from those of similar age elsewhere. It consists mainly of hypsilophodontids, small ornithopods that are known abundantly in the Early Cretaceous of Europe and North America. Other elements of the fauna consist of an allosaurid, crocodilians, pterosaurs, turtles, amphibians, and fishes. The only hints that the dinosaurs were in any way adapted to their polar conditions is that there were no large dinosaurs in the fauna, and one of them seems to have had extra-strong eyesight, perhaps, hinted the discoverers, an adaptation to living in winter darkness. (Or perhaps not.)

The commonest dinosaurs from the Early Cretaceous of Victoria were hypsilophodontids, small active herbivores (above), as recent excavations in the 1990s have shown. Large dinosaurs were generally absent. These are part of the largest dinosaur fauna from southern polar regions, although specimens have also been discovered at several localities in Antarctica, and in New Zealand (below).

17 Plesiosaur

18 Prosauropod

19 Pterosaur

20 Sauropod

21 Stegosaur

22 Tritylodontid

23 Troodontid

24 Turtle

25 Carnosaur footprint

26 Iguanodont footprint

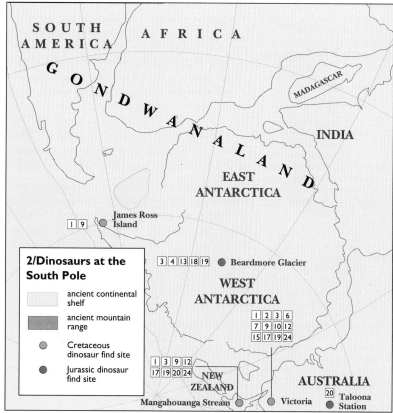

SOUTH AMERICA

AFRICA

GONDWANALAND

MADAGASCAR

INDIA

EAST ANTARCTICA

1 9 · James Ross Island

3 4 13 18 19 ● Beardmore Glacier

WEST ANTARCTICA

1 2 3 6
7 9 10 12
15 17 19 24

1 3 9 12
17 19 20 24

2/Dinosaurs at the South Pole

ancient continental shelf

ancient mountain range

○ Cretaceous dinosaur find site

● Jurassic dinosaur find site

NEW ZEALAND

AUSTRALIA

20 Taloona Station

Mangahouanga Stream ● Victoria ●

Early Cretaceous Dinosaurs of Africa

Dinosaurs have been known from the Early Cretaceous of North Africa for some time, but recent discoveries have shown the true potential of the area.

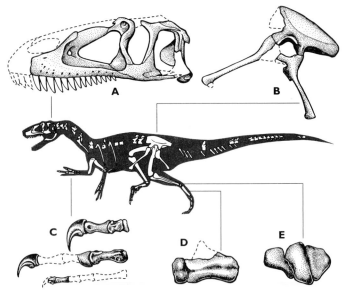

Dinosaurs have been found in a number of isolated localities in rocks of Early Cretaceous age, but in most localities, apart from those in the Wealden of north-west Europe (see pages 118–119), the fossils are limited. Until recently, dinosaurs were incompletely known from the Early Cretaceous of Africa.

The first discoveries of Early and Mid Cretaceous dinosaurs in Africa resulted from expeditions that had originally been searching for Tertiary mammals. The German palaeontologist Ernst Stromer von Reichenbach visited Egypt many times from 1901 to 1911, and discovered a dinosaur graveyard at the Baharija oasis, in the western desert of Egypt south of Fayum. In a rich fauna the prize discovery was the theropod dinosaur *Spinosaurus*.

The latest finds from North Africa have included the new genus Afrovenator (above), reported by Paul Sereno in 1994. This theropod is large, some 8–9 metres long, and an agile predator that fed on sauropod dinosaurs that lived with it in Niger during the Early Cretaceous. The skull (A) was nearly complete, and it showed close affinities with Torvosaurus from North America. A particular feature of torvosauroids is the long tooth-bearing maxillary bone. The pelvis (B) is lightly built, with long slender pubis and ischium bones below. The hand (C) had three fingers, and it was designed for grasping prey items, while the ankle (D) and upper foot bones (E) show features typical of other large theropods in the torvosauroid group. Paul Sereno's expedition in 1995 turned up a complete skull of Carcharodontosaurus, perhaps the largest meat-eater of all time.

Spinosaurus looked like any of a number of theropods, such as *Allosaurus*, except that it had tall spines on its vertebrae, the longest ones reaching 2 metres in length. It seems likely that these spines were used to support a sail on the back, something like the sails of the much more ancient, and unrelated, pelycosaur mammal-like reptiles. If *Spinosaurus* had a sail, this could have been used partly for signalling and sexual display, but it is such a substantial feature that it probably functioned in thermoregulation too. On cool mornings, *Spinosaurus* would stand sideways to the sun, and allow the heat from the rays to seep in and warm the blood. When the dinosaur overheated at midday, it could shelter in the shade, and pump blood through the sail to radiate heat, thus cooling the body.

Amazingly, sails of this kind have been found also in other North African dinosaurs. French explorers visited an extensive series of Early Cretaceous deposits in the western part of the Sahara, the 'continental intercalaire' and Elrhaz Formation of Morocco, Algeria, Tunisia, Libya, Mali, and Niger from the 1950s to the 1970s. Here, in Niger, Philippe Taquet found a nearly complete skeleton of the ornithopod *Ouranosaurus* in the 1970s. *Ouranosaurus* turned out to be very similar to *Iguanodon*, a typical Early Cretaceous ornithopod from Europe, as well as China and North America (see pages 118–119), but the new African dinosaur also had a sail on its back.

Dramatic new discoveries from North Africa have been reported more recently by Paul Sereno, who visited Niger in 1993 and 1994, and Morocco in 1995. His 1993 expedition yielded a sauropod skeleton, and a new theropod, *Afrovenator*, which turned out to be most similar to the North American *Torvosaurus*. This is tantalising evidence for land links between the two continents, even after they were thought to have separated.

Spinosaurus, from the Mid Cretaceous of Egypt, the first dinosaur to be reported from North Africa. This predatory dinosaur is shown feeding on the herbivore Ouranosaurus, *also from North Africa, a close relative of* Iguanodon. *The dorsal sail in both forms may have been used in temperature control. Note that Egypt lay very close to the Cretaceous equator (below).*

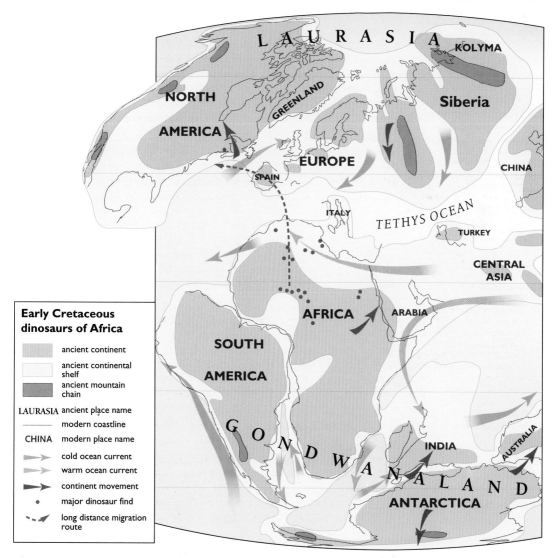

Early Cretaceous dinosaurs of Africa

	ancient continent
	ancient continental shelf
	ancient mountain chain
LAURASIA	ancient place name
--------	modern coastline
CHINA	modern place name
	cold ocean current
	warm ocean current
	continent movement
•	major dinosaur find
--- ⌁	long distance migration route

LAURASIA
KOLYMA
NORTH AMERICA
GREENLAND
Siberia
EUROPE
CHINA
SPAIN
ITALY
TETHYS OCEAN
TURKEY
CENTRAL ASIA
AFRICA
ARABIA
SOUTH AMERICA
INDIA
AUSTRALIA
GONDWANALAND
ANTARCTICA

Dinosaurs of Gondwanaland

Cretaceous dinosaurs of the southern hemisphere were like northern dinosaurs in many ways, but recent discoveries have included some surprises.

By Mid and Late Cretaceous times, Gondwanaland had broken up, partly as a result of continental drift, and partly as a result of flooding by rising sea levels (see pages 116–117). The dinosaurs from South America, Africa, India, Antarctica, and Australia share many similarities—after all, their faunas of dinosaurs originated long before, when Gondwanaland was still united (see pages 70–71, 96–97). Indeed, the Cretaceous Gondwanan dinosaurs were still like those of the northern hemisphere in many ways, but there were already some major differences, and these differences are now being used in attempts to reconstruct ancient continental movements in detail.

Skull of Carnotaurus, *an abelisaurid theropod from Argentina.*

The differences became clear when large-scale dinosaur collecting began early in the 20th century. The main work was initially in South America. In 1882, an official, Commandante Buratovich, found some very large bones in Upper Cretaceous sediments near the city of Neuquén in Patagonia, southern Argentina. These fossils were sent to Florentin Ameghino, a renowned Argentinian palaeontologist, who identified them as those of dinosaurs. His brother, Carlos, set off for the field, and found more specimens, and other collectors began amassing bones, some of which were described by the Ameghino brothers, and others of which were sent to experts in England and Germany for their opinions. In 1921 and 1922, the Museum of La Plata mounted major excavations in the Neuquén region under the direction of Santiago Roth, a man who had found dinosaur bones in the vicinity in the 1880s. Large collections were made and sent to the museum in La Plata.

The most striking new finds were the bones of huge sauropod dinosaurs, called the titanosaurs. By 1922, the Ameghino brothers were dead, and there was no Argentinian palaeontologist who could take on the task of describing

I/Cretaceous Gondwanaland

▨	ancient continent
☐	ancient continental shelf
▨	ancient mountain chain
BALTICA	ancient place name
——	modern coastline
CHINA	modern place name
➤	cold ocean current
➤	warm ocean current
➤	continent movement
○	titanosaur site
●	abelisaurid site

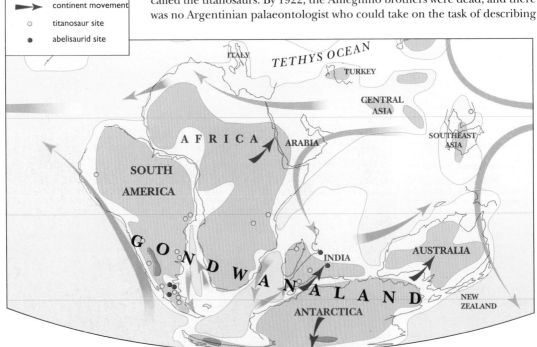

2/ The Passage of India, Cretaceous to Eocene.

Late Cretaceous migration

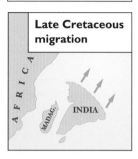

Paleocene isolation

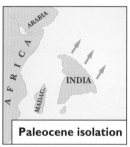

Early Eocene docking

India existed as an island during the Cretaceous (above), and joined Asia only in the Eocene—Some largely Gondwanan dinosaurs existed in India and in other Southern Continents (left). The Late Cretaceous dinosaurs of Argentina include some mainly Gondwanan forms such as the giant sauropod Titanosaurus *(right), and the abelisaurid theropod* Carnotaurus. *Sauropods were nearly extinct in the rest of the world, but the titanosaurids were dominant plant-eaters in Gondwanaland, taking the role of the ornithopods.*

the astonishing new collections. The authorities of the Museum of La Plata invited Friedrich von Huene to do the job. He was still a young man, but had built a strong international reputation for his work on the dinosaurs of Germany and other parts of Europe (see pages 76–77). He arrived in Argentina, and worked with his usual vigour, both in the museum, and in the field. The result was a vast publication, *Los Saurisquios y Ornitisquios del Cretáceo Argentino*, in which von Huene described *Titanosaurus* in detail, as well as the other Neuquén dinosaurs.

Von Huene's publications about the Argentinian dinosaurs introduced the world to the titanosaurs, a new sauropod group, and hinted that the southern continents might have housed some unusual dinosaur groups that were not known from Europe or North America. Titanosaurs were also known from India, and they were later found in Brazil, Malawi, Madagascar, and Laos, as well as in Romania (see pages 128–129) and the southern United States. The group was truly Gondwanan, but with a couple of finds in the southern parts of Laurasia, north of the Tethys Sea. Titanosaurids were large sauropods, and some like *Argentinosaurus*, were as large as the largest diplodocids of the Late Jurassic (see pages 102–103). Recent discoveries of a form called *Saltasaurus* have shown that titanosaurids had bony plates set in the skin, a kind of chain-mail body armour.

Further recent discoveries in South America have confirmed this pattern. José Bonaparte (see pages 96–97) recently described an unusual new meat-eating dinosaur, *Carnotaurus*. The skull is deep and there is a pair of horn-like projections above the eye sockets. He realised this was an entirely new dinosaur type, and included it in a new family, the Abelisauridae. Further abelisaurids came to light in Argentina, as well as in India, and, unexpectedly, in southern France. The abelisaurids are another Gondwanan family, but evidently a few animals were somehow able to cross the widening Tethys Sea to reach southern Europe.

Dinosaurs of Mongolia and China

When dinosaurs were first reported from Mongolia and China, they caused a sensation. Dramatic new discoveries are still being made.

Roy Chapman Andrews (above) and his assistant, George Olsen, excavating a nest of Protoceratops *eggs at Bayn-Dzak, in the Gobi Desert, Mongolia, in the 1920s. This was one of the remotest parts of the world to visit, and the expedition was a logistical nightmare. The discovery of nests of dinosaur eggs caused great excitement. The nests were built as hollows in the ground, and the elongate eggs had been laid in several concentric circles, making a total of about twenty per nest.*

The first dinosaur discoveries in Mongolia were made in the 1920s by teams from the American Museum of Natural History. But the director of the museum, Henry Fairfield Osborn, was not looking for dinosaurs.

The purpose of the expedition was to find fossil humans. Osborn, and his colleague Roy Chapman Andrews, believed that humans had arisen in central Asia, and they thought that Mongolia might hold the key. The country was little known in the West, but Vladimir A. Obruchev, a distinguished Russian palaeontologist, had found a rhinoceros tooth in Mongolia in 1892. This suggested that rocks of the right age to hold fossil humans might be there. Andrews' caravan of vehicles and camels set off from China in early 1922, crossed the border into Mongolia, and headed north towards Ulan Baatar. Andrews, and the chief palaeontologist Walter Granger made discoveries early on, first at a site called Iren Dabasu, on the road to Ulan Baatar, where they found Cretaceous mammals and dinosaurs. They visited Ulan Baatar briefly, and then set off west into the Gobi Desert.

After some weeks of exploring the Gobi Desert, and a few discoveries, the expedition turned east again, heading back to China. One day, the vehicles drew up on the edge of a large eroded basin formed in red sandstones, a site they named 'Flaming Cliffs', and now officially called Bayn-Dzak. The collectors found abundant dinosaur bones and eggs, but they then had to head home. The expedition returned in 1923, and they were able to spend adequate time at Iren Dabasu and at Bayn Dzak. They collected bones of several extraordinary new dinosaurs. There were dozens of specimens of the small ceratopsian *Protoceratops*, and, most dramatically of all, these were associated with several nests containing elongate eggs arranged in neat circles. The collections of *Protoceratops* with their nests, created a sensation. Dinosaur eggs had been found as isolated remains in the Late Cretaceous of southern France in the 19th century, but these were the first complete nests.

The Americans returned to the Gobi Desert in 1925, and during their three expeditions they amassed large numbers of specimens. Apart from *Protoceratops*, they found small theropods such as *Saurornithoides* and *Oviraptor*. Later expeditions to Mongolia were mounted by the Russians in 1946, 1947, and 1949, and they found new dinosaurs, especially *Tarbosaurus*, a relative of the North American Tyrannosaurus (see pages 132–133) and the hadrosaur Saurolophus, also a North American form. Polish expeditions operated in Mongolia in the 1960s, and the American Museum of Natural History began a second series of expeditions in the 1990s, all of which have been crowned with astonishing new discoveries.

Dinosaurs were found in China in the 1920s, and some of the most important finds are Jurassic in age (see pages 94–95). Dramatic discoveries have been made in the Cretaceous also by Chinese scientists since the 1970s, and their dinosaurs are in many ways like the Mongolian ones.

The dinosaurs of Late Cretaceous Asia (right) include the extraordinary hadrosaur Tsintaosaurus from China. It has a vertically-standing crest on its head. The crest is made from the nasal and frontal bones, which normally lie in front of the eyes. In the hadrosaurs, or duckbilled dinosaurs, a Late Cretaceous group of ornithopods, these two skull bones have moved back and they extend into bizarre and complex crests, which were probably used as visual signals, and perhaps audible ones too: the nasal passages ran through the crests, and when Tsintaosaurus snorted, it made a bellowing noise. The other hadrosaur, Saurolophus, has a lower crest, and it is very similar to North American forms.

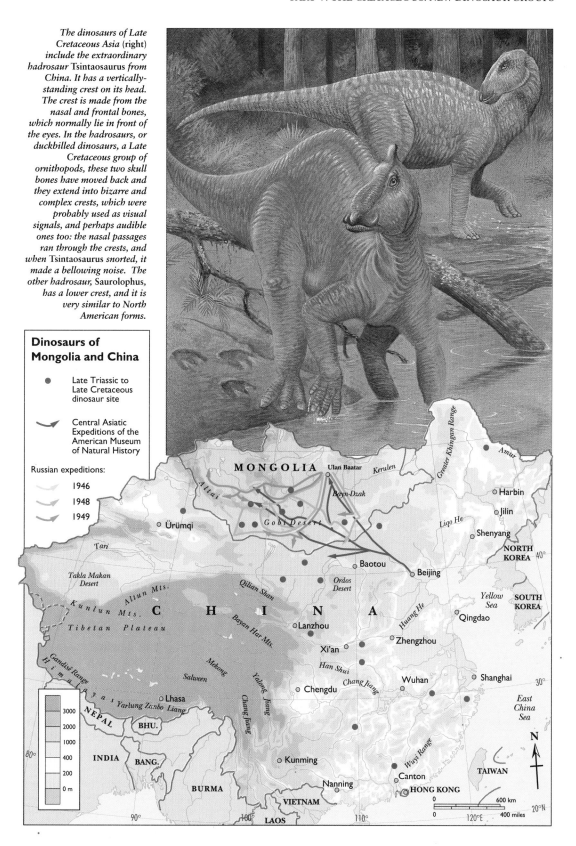

Dinosaurs of Mongolia and China

- Late Triassic to Late Cretaceous dinosaur site

↝ Central Asiatic Expeditions of the American Museum of Natural History

Russian expeditions:

↝ 1946

↝ 1948

↝ 1949

127

Dwarf dinosaurs of Romania

In the Late Cretaceous, much of eastern Europe was an archipelago of islands. Dinosaurs on the islands became dwarfs.

During the Cretaceous, the major modern continents were moving towards their present positions, but continents north and south were separated by the Tethys Ocean. This ran from western Europe and north Africa to China and central Asia. In the region of eastern Europe and the Middle East, the Tethys Ocean was particularly broad, and only a few volcanic islands broke through. After the Cretaceous, the Ocean closed up, and land was formed.

One of the eastern European islands, Hateg (pronounced Hat-zeg) Island, has produced a rich dinosaurian fauna. Hateg is located now in southern Romania. During the Late Cretaceous, the island measured about 7500 km², and it lay in an archipelago some 200 to 300 km from the nearest main land mass, in northern Europe. The sediments consist of river deposits with tropical soils, and here and there, ashes, testifying to continuing volcanic activity and uplift of the island.

Dinosaur bones were first found by Ilona Nopcsa, on her family estates in the Hateg area in the 1890s. She showed the bones to her brother, Baron Franz Nopcsa, and he began his palaeontological career by describing these specimens, and further collections that were excavated more systematically. Further excavations since 1970 have increased our knowledge of the Hateg dinosaur fauna. There are about ten dinosaur species, as well as freshwater fishes, amphibians, a crocodile, a turtle, a pterosaur (flying reptile), and a mammal. The dinosaurs include small and large theropods, a titanosaurid sauropod, an ankylosaur (*Struthiosaurus*), and several ornithopods (*Telmatosaurus, Rhabdodon*). Dinosaur nests with eggs have also been found.

This reasonably diverse dinosaur fauna includes forms fairly typical of the Late Cretaceous of other parts of the world. Several of the Hateg dinosaurs (*Telmatosaurus, Struthiosaurus*) are, however, unusually primitive, and this suggests that their ancestors reached the island earlier in the Cretaceous, and they survived there long after their relatives elsewhere had died out. Most of the Hateg dinosaurs show links with typical Euramerican faunas but the titanosaurid sauropod is an unusual element, representing a group that is best known in southern continents, particularly South America, Africa, and India (see pages 124–125). Isolated titanosaurid specimens have been reported from southern France, so the group clearly reached Europe, perhaps via north Africa and the Iberian peninsula.

The best-known of the Hateg dinosaurs, Telmatosaurus, was a primitive hadrosaur, or duck-billed dinosaur. It was about one-third the length, and one-tenth the weight of its closest relatives from elsewhere.

More unusual is the relatively small size of the Hateg dinosaurs. *Telmatosaurus* is about 5 m long, and it probably weighed about 500 kg, large enough perhaps, but only 10% of the weight of its relatives from elsewhere. The Romanian skeletons represent adults, as shown by the fusion of their bones, and hence this is a dramatic example of dwarfing on an island. On islands today, land animals may become either smaller or larger. The size change is caused by modified ecological situations on islands. Island faunas usually contain fewer species than the faunas on the nearest mainlands, and normal competitor species may be absent. Hence, modern birds and mammals can broaden their range of diets, and their sizes may change as they modify rapidly to changed circumstances. The Hateg dinosaurs may have become smaller in order to exploit feeding possibilities left vacant by the absence of other dinosaur species.

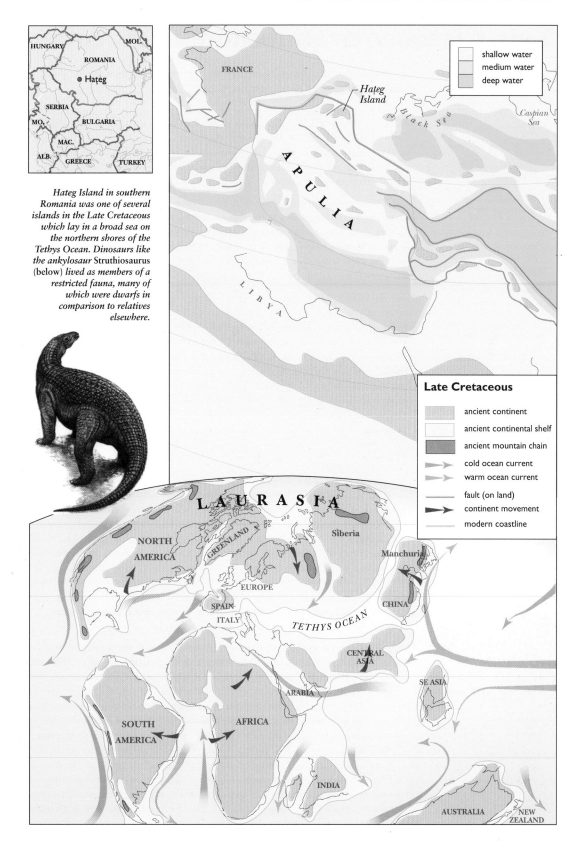

HUNGARY

MOL.

ROMANIA

● Hateg

SERBIA

MO.

BULGARIA

MAC.

ALB.   GREECE

TURKEY

FRANCE

Hateg
Island

shallow water
medium water
deep water

Black Sea

Caspian
Sea

A P U L I A

L I B Y A

*Hateg Island in southern
Romania was one of several
islands in the Late Cretaceous
which lay in a broad sea on
the northern shores of the
Tethys Ocean. Dinosaurs like
the ankylosaur Struthiosaurus
(below) lived as members of a
restricted fauna, many of
which were dwarfs in
comparison to relatives
elsewhere.*

Late Cretaceous

ancient continent

ancient continental shelf

ancient mountain chain

cold ocean current

warm ocean current

fault (on land)

continent movement

modern coastline

L A U R A S I A

Siberia

NORTH
AMERICA

GREENLAND

Manchuria

EUROPE

CHINA

SPAIN

ITALY

TETHYS OCEAN

CENTRAL
ASIA

SE ASIA

ARABIA

SOUTH
AMERICA

AFRICA

INDIA

AUSTRALIA

NEW
ZEALAND

Late Cretaceous Dinosaurs of N. America

The dinosaurs of the last 20 million years of the Cretaceous in North America are some of the richest and most diverse in the world.

Reconstruction of an embryo Maiasaura *in its egg (above). Numerous nests containing eggs have been found at and around Egg Mountain, Montana, since the 1970s, and these include a few rare eggs with enclosed embryos. Painstaking work by preparators at the Museum of the Rockies has revealed the tiny bones inside the shell, and museum artists have been able to reconstruct the tiny embryo, curled up and ready to hatch out.*

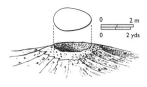

Maisaura nests are 2 metres across, and formed by hollows scooped in the earth (above). Eggs were laid in the hollow, and this was covered with leaves to provide warmth as they rotted (like a compost heap). Vertical sections through the sediments of Egg Mountain, Montana (below), show how nests were produced at several different levels, suggesting that Maiasaura *mothers returned year after year to lay their eggs.*

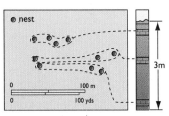

The last phases of the Cretaceous period show the history of the last dinosaurs, and the record in North America is one of the best in the world. The last 20 Myr of the Cretaceous is divided into the Santonian, Campanian, and Maastrichtain stages, and dinosaurs have been found abundantly at all levels.

Late Cretaceous dinosaurs were found first in North America on the east coast, when Leidy described *Hadrosaurus* from New Jersey in 1858 (see pages 24–25), and huge collections were made by teams working for Marsh and Cope. The story really picked up when palaeontologists began exploring the rich dinosaur beds of Alberta, Canada in the early 20th century. About 1910, two groups set out along the Red Deer River to collect dinosaur skeletons.

The Red Deer River cuts deep canyons through the Late Cretaceous sediments of southern Alberta. Climatic conditions are desert-like for most of the year, and there is very little vegetation other than scrubby bushes. Every year, there are torrential rains, and huge rivers cut through the canyon walls, creating classic badland scenery (bad for farmers; good for fossil-hunters). The two collecting teams were led by Barnum Brown, acting for the American Museum of Natural History, and Charles H. Sternberg, working for several institutions, but especially the Geological Survey and the National Museum of Canada, and the Royal Ontario Museum.

Barnum Brown invented a new collecting technique. He built a large wooden barge, and floated downstream in 1910 and 1911, tying up here and there, and venturing up side canyons looking for shards of bone. When he found a good prospect, he and his team excavated the bones, and loaded the plaster packages on to the barge. They were later offloaded and sent east to New York. The Canadian Government was aware of Brown's discoveries, and they decided to secure some dinosaurs for their own fledgling museums. They hired Charles Sternberg, a commercial collector, who operated at times with the help of some or all of his sons, George, Charles, and Levi. The Sternbergs set off for the Red Deer River in 1912, and they found rich dinosaur beds around Drumheller and Steveville. Both teams returned in 1913, and trips continued until 1917. During this time, they found dozens of complete skeletons of dinosaurs at various levels in the Late Cretaceous.

Similar sequences of Late Cretaceous rocks occur across the border in the United States, in Montana and the Dakotas. Long-term collecting by Jack Horner and his colleagues in the Montana Badlands since the 1970s has turned up yet another series of dramatic discoveries. He located a dinosaur nesting ground where hadrosaurs, or duckbilled ornithopods, had apparently returned year after year to lay their eggs. Horner found numerous nests on an elevated patch of ground, and by digging through the sediments of the site, which he named Egg Mountain, he found that the same hadrosaur species had built nests here time after time. The nests were shallow hollows scraped in the ground, and the eggs were apparently tended by the parents and older offspring, since he found skeletons of all ages groups associated with the nests. He suggests that dinosaurs showed intelligent parental care, and he named the new hadrosaur *Maiasaura*, or 'good mother reptile'.

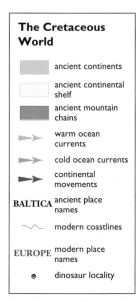

The Cretaceous World

- ancient continents
- ancient continental shelf
- ancient mountain chains
- warm ocean currents
- cold ocean currents
- continental movements

BALTICA — ancient place names

- modern coastlines

EUROPE — modern place names

- dinosaur locality

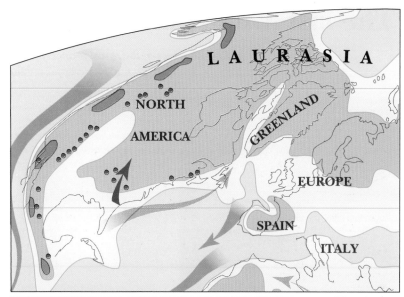

Late Cretaceous dinosaur sites are widespread in North America (above). Abundant finds have been made in badland sediments, such as these on the South Saskatchewan River in Alberta.

The last American dinosaurs

The last dinosaurs in the world include **Tyrannosaurus, Triceratops,** *and the hadrosaurs, all highly successful animals living in a modern world.*

Collecting dinosaurs in the latest Cretaceous of Montana (above). *Here, the collectors are cleaning a dinosaur arm bone before covering it with sheets of sackcloth soaked in plaster. Bones are scattered all over this site, and the whole area has to be cleared and mapped carefully so nothing is lost.*

Dinosaurs of the latest Cretaceous of midwestern North America (opposite) *included hadrosaurs such as* Corythosaurus *with its rounded plate-like crest* (middle right). Triceratops, *a horned dinosaur* (bottom right), *was also common at the time, one of the last-surviving dinosaurs. The great flesh-eater* Tyrannosaurus (left) *looks hungrily at these plant eaters. These dinosaurs lived in landscapes that looked rather modern. There were flowering plants like roses and magnolias, small mammals, modern-looking seabirds, frogs, lizards, and snakes. To our eyes, perhaps only the dinosaurs strike an antiquarian note.*

The extinction of the dinosaurs has always been a question of interest to palaeontologists, and to the public. Early in the 20th century, the most popular notion was that the dinosaurs were exhausted, mere remnants of their former selves, and that they became extinct simply because they had served their time. Specific evidence was found in the fact that many Late Cretaceous dinosaurs, such as the ceratopsians and hadrosaurs, had horns and crests on their heads. These were said to show wild patterns of evolution that merely produced useless structures that encumbered the animals.

The idea that the dinosaurs were doomed to extinction was called 'racial senility'. Extinction was seen to be inevitable at a certain point in the evolution of a group, and dinosaurs were treated as a classic example. The image of dinosaurs as unsuccessful animals, too big and inefficient to survive is now unsustainable, since dinosaurs were clearly some of the most successful animals ever (see page 8). The horns and crests of Late Cretaceous dinosaurs were not useless, there is no evidence for racial senility.

In fact, the last dinosaurs were some of the most varied ever seen. Dominant by far were the hadrosaurs, or duckbilled dinosaurs, a major flowering of the ornithopod group which included, earlier in the Cretaceous, forms like *Iguanodon* (see pages 118–119). A visitor to the Late Cretaceous of North America, or indeed Mongolia or China (see pages 126–127), would have seen vast herds of hadrosaurs. The herds would have contained animals of similar size, and with similar body shapes, but the headgear was wildly different. Some had plate-like crests, others long spines, and others spikes. The crests were made from the bones associated with the nasal passages, and they were once thought either to have had no function, but current work (see pages 112–114) shows they were visual and audible signalling devices, used by members of each species to recognise each other, for parents to find their young, and for ritual use before mating.

Associated with the hadrosaurs were ankylosaurs, ceratopsians, and various theropods. The ankylosaurs of the latest Cretaceous, such as *Ankylosaurus* (see page 135), were huge heavily armoured plant-eaters that were probably virtually free from predation. The ceratopsians, such as *Triceratops*, generally had horns on their snouts and faces, as well as a bony shield that extended from the back of the skull to protect the neck. It is likely that the ceratopsians used their horns to defend the herd, by facing outwards towards a predator. Latest Cretaceous theropods include the small highly intelligent troodontids, the larger ostrich-like ornithomimids, and the largest carnivores of all, the tyrannosaurids. *Tyrannosaurus rex* (see page, 111–113) is one of the most evocative dinosaurs, a giant flesh-eater that fed on all of the plant-eaters of its day, using its vast jaws and 1-metre gape to good effect.

Bones of *Triceratops* and *Tyrannosaurus* are found within a few metres of the last sediments of the Cretaceous in Montana, so these highly successful dinosaurs were some of the last ever to breath on the earth.

The Extinction of the Dinosaurs

Dinosaurs became extinct 65 million years ago, and this seems to have been triggered by the impact of an asteroid in the Caribbean.

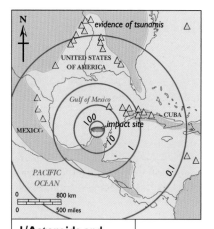

1/Asteroids and iridium

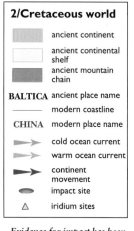

lines of equal thickness of material ejected from crater

△ main iridium site

The Chicxulub impact site in the Yucatán peninsula.

2/Cretaceous world

	ancient continent
	ancient continental shelf
	ancient mountain chain
BALTICA	ancient place name
———	modern coastline
CHINA	modern place name
⇒	cold ocean current
⇒	warm ocean current
➤	continent movement
●	impact site
△	iridium sites

Evidence for impact has been found worldwide (right). Iridium enhancements, a sure marker of the arrival of meteorites, have been found at 200 localities, as has shocked quartz, evidence for high-pressure impact. These materials arrived as fall-out, and they are found in sediments deposited both on land and in the sea.

It wasn't only the dinosaurs that disappeared 65 million years ago at the Cretaceous–Tertiary (KT) boundary. In addition, the pterosaurs died out, as well as several families of birds and marsupial mammals. In many other important Mesozoic groups disappeared at about the same time, such as the plesiosaurs, mosasaurs, ammonites, belemnites, rudist and trigoniid bivalves, and various plankton groups in the sea. Most plants and many animals, however, were unaffected, such as gastropods, most bivalves, fishes, amphibians, turtles, lizards, and placental mammals. It is hard to separate the survivors and non-survivors into simple ecological categories.

Research on the KT event increased after 1980, when Luis Alvarez of the University of California, and colleagues, published their view that the extinctions had been caused by the impact of a 10km–diameter asteroid on the Earth. The impact caused massive extinctions by throwing up a vast dust cloud which blocked out the sun and prevented photosynthesis, and hence plants died off, followed by herbivores, and then carnivores. The dust cloud also prevented the sun's heat reaching the earth, and there was a short freezing episode.

There are three key pieces of evidence for the impact hypothesis, an iridium anomaly worldwide at the KT boundary, and associated shocked quartz and glassy spherules. Iridium is a platinum-group element that is rare on the Earth's crust, and reaches the Earth in meteorites. At the KT boundary, that

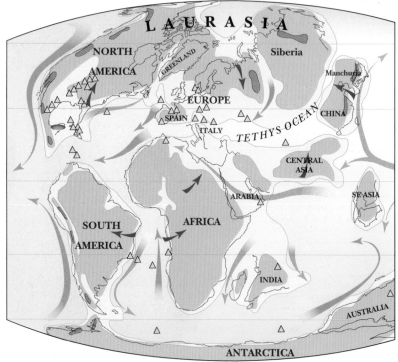

Some geologists argue that the Deccan traps of India were critical in causing the KT extinction (below). 65 Myr ago, huge volumes of volcanic basalt were erupted over northern India, and these would have caused profound environmental changes over a wide area.

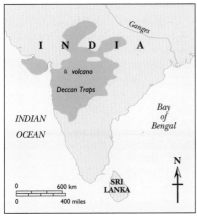

Relative abundance of dinosaurs and mammals in the last 10 million years of the Cretaceous in the Hell Creek Basin of Montana

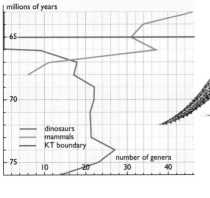

Some evidence suggests that the dinosaurs of the American Midwest experienced long-term decline. Plots of dinosaur diversity (above) show an apparent reduction in numbers of species lasting for the last 5 Myr. of the Cretaceous. At the same time, mammal diversity increased. Some of the last dinosaurs were the heavily armoured ankylosaurs (right), including the nodosaurs (top; left) and the ankilosaurids, with their bony tail clubs.

rate increased dramatically. Many localities have also yielded shocked quartz, grains of quartz bearing crisscrossing lines produced by the pressure of an impact, as well as glassy spherules produced by melting.

The second school of thought concerning the KT events has focused on explaining the iridium spike, the shocked quartz, and the glassy spherules by volcanic activity. The Deccan Traps in India represent a vast outpouring of lava which occurred over the 2-3 Myr. spanning the KT boundary. Petrologists and geochemists argue, however, that the shocked quartz and iridium spike could not be produced by any known kind of volcano.

The main alternative to the extraterrestrial catastrophist explanation for the KT mass extinction is the gradualist model, which sees declines in many groups caused by long-term climatic changes in which subtropical lush dinosaurian habitats gave way to strongly seasonal temperate conifer-dominated mammalian habitats. This model has been challenged on the basis of problems in exact correlation of the isolated mammal faunas. The gradualist scenario has been extended to cover all aspects of the KT events on land and in the sea, with evidence from the gradual declines of many groups through the Late Cretaceous. Climatic changes on land are linked to changes in sea level and in the area of warm shallow-water seas.

The impact model has been strengthened by the discovery, in 1990, of the impact site, the Chicxulub Crater in Mexico. This is big enough (200 km diameter) for a 10km asteroid, and there is strong evidence for impact fallout and tsunamis (tidal waves) all round the Proto-Caribbean. Perhaps the smoking gun has been found, but palaeontologists still have to explain how the impact actually caused the selective extinctions.

Further Reading

GENERAL BOOKS

There are numerous dinosaur books aimed mainly at children: probably 100 or more are published each year in the English language. I have selected here some books aimed at a more adult, or informed, audience. The 'bible' for dinosaur palaeontologists is Weishampel *et al.* (1990), Norman (1985) is an excellent book with superb colour paintings by John Sibbick, Benton (1989) is a highly illustrated account of how dinosaurs are excavated and studied, and Fastovsky and Weishampel (1996) is the best textbook.

Benton, M.J., *On the Trail of the Dinosaurs*, Crescent Books, New York, and Kingfisher, London, 1989.

Fastovsky, D.E. and Weishampel, D.B. *The Evolution and Extinction of the Dinosaurs*, Cambridge University Press, Cambridge, 1996.

Norman, D.B., *The Illustrated Encyclopedia of Dinosaurs*, Salamander, London, 1985.

Weishampel, D.B., Dodson, P., and Osmólska, H, eds, *The Dinosauria*, University of California Press, Berkeley, 1990.

SPECIALIST BOOKS

There are many books that concentrate on specific aspects of the dinosaurs, and on the history of their collection and study. In addition, some books on related aspects of palaeontology and earth history may provide useful background reading (Benton, 1997; Benton and Harper, 1997; Stanley, 1986). Colbert (1968) is an excellent outline of the history of dinosaur collecting.

Alexander, R.McN, *The Dynamics of Dinosaurs and Other Extinct Giants*, Columbia University Press, New York, 1989.

Archibald, J.D., *Dinosaur Extinction and the End of an Era: What the Fossils Say*, Columbia University Press, New York, 1996.

Bakker, R.T., *Dinosaur Heresies*, William Morrow, New York and Penguin, London, 1986.

Benton, M.J., *The Reign of the Reptiles*, Crescent Books New York and Kingfisher, London, 1990.

Benton, M.J., *The Rise of the Mammals*, Crescent Books, New York, Apple Tree Press, London, 1991.

Benton, M.J., *Vertebrate Palaeontology*, 2nd edn, Chapman & Hall, London, 1997.

Benton, M.J. and Harper, D.A.T., *Basic Palaeontology*, Addison-Wesley Longman, London, 1997.

Benton, M.J., Kurochkin, E.N., Shishkin, M.A., and Unwin, D.M., eds, *The Age of Dinosaurs in Russia and Mongolia*, Cambridge University Press, Cambridge, 1997.

Carpenter, K. and Currie, P.J., *Dinosaur Systematics*, Cambridge University Press, New York, 1990.

Colbert, E.H., *Men And Dinosaurs*, Dutton, New York and Duckworth, London, 1968.

Czerkas, S.J. and Olson, E.C., eds, *Dinosaurs Past and Present*, 2 vols, Natural History Museum of Los Angeles County, Los Angeles, 1987.

Desmond, A., *The Hot-Blooded Dinosaurs*, Blond and Briggs, London and Dial Press, New York, 1975.

Feduccia, A., *The Age of Birds*, Harvard University Press, Cambridge, Mass., 1980.

Gillette, D.D. and Lockley, M.G., eds, *Dinosaur Tracks and Traces*, Cambridge University Press, Cambridge, 1990.

Hecht, M.K., Ostrom, J.H., Viohl, G., and Wellnhofer, P., eds, *The Beginnings of Birds*, Freunde de Jura-Museums Eichstätt, Willbaldsburg, 1984.

Horner, J.R. and Lessem, D., *The Complete T. rex*, Simon & Schuster, New York, 1993.

Kielan-Jaworowska, Z., *Hunting for Dinosaurs*, MIT Press, Cambridge, Mass., 1969.

Lessem, D., *Kings of Creation: How a New Breed of Scientists is Revolutionizing our Understanding of Dinosaurs*, Simon & Schuster, New York, 1992.

Russell, D.A., *The Dinosaurs of North America*, University of Toronto Press, 1989.

Schultze, H.-P. and Trueb, L., eds, *Origins of the Higher Groups of Tetrapods: Controversy and Consensus*, Cornell University Press, Ithaca, NY, 1991.

Sharpton, V.L. and Ward, P.D., eds, *Global Catastrophes in Earth History*, Geological Society of America, Boulder, Colo., 1990.

Stanley, S.M., *Earth and Life through Time*, W.H. Freeman, San Francisco, 1986.

Sternberg, C.H., *Hunting Dinosaurs in the Bad Lands of the Red Deer River, Alberta, Canada*, NeWest Press, Edmonton, Alberta, 1985.

Thomas, R.D.K. and Olson, E.C., eds, *A Cold Look at the Warm-Blooded Dinosaurs*, Westview Press, Boulder, Colo., 1980.

Wellnhofer, P., *The Illustrated History of Pterosaurs*, Salamander Books, London, and Crescent Books, New York, 1991.

Wilford, J.N., *The Riddle of the Dinosaur*, Knopf, New York, 1985.

Index

References shown in **bold** are maps or pictures. Quotes are in *italics.*

Acknowledgements

Picture Credits

Front Cover Illustrations

(clockwise from top left)

A scene from a Carboniferous forest. One of the first reptiles is shown. Huge lycopsid trees dominated the tropical swamps, insects and reptiles thrived in the damp warm conditions.

Saltasaurus

Ankylosaurus

Apatosaurus

Woodlands of 70 million years ago showing various flowering plant families.

Triceratops

Diplodocus

Iguanodon

Internal

Michael Benton: 30, 32, 34, 131, 132

Professor E.H. Colbert, Museum of Northern Arizona: 76b

Permission of Dr. Bernhard Krebs, Lehrstühl für Paläontologie der Freien Universität, Berlin, and Dr. Werner Janensch, Berlin: 98

Trustees of the National Museums of Scotland: 92

The Natural History Museum, London: 13tl, 14, 24tl, 26, 126

Anonymous Sources: 15, 17, 22, 28tl, 31, 76tl,

FOR SWANSTON PUBLISHING LIMITED

Concept:
Malcolm Swanston

Picture Research:
Charlotte Taylor

Editorial:
Stephen Haddelsey

Production:
Andrea Fairbrass
Barry Haslam

Illustration:
Ralph Orme
Steve Roberts
Peter Smith

Separations:
Central Systems,
Nottingham.

Cartography:
Andrea Fairbrass
Elsa Gibert
Kevin Panton

Index:
Jean Cox
Barry Haslam

Typesetting:
Andrea Fairbrass
Charlotte Taylor